Student Solutions Manual

Phyllis Barnidge for Laurel Technical Services

BASIC MATHEMATICS

Barbara Poole

North Seattle Community College

Editorial/production supervision: ***Amy Jolin***
Editor in Chief: ***Tim Bozik***
Acquistion editor: ***Jerome Grant***
Supplements acquisitions editor: ***Audra Walsh***
Production coordianator: ***Trudy Pisciotti***

Simon & Schuster / A Viacom Company
Upper Saddle River, New Jersey 07458

Printed in the United States of America
10 9 8 7 6

ISBN 0-13-458050-8

Prentice-Hall International (UK) Limited, *London*
Prentice-Hall of Australia Pty. Limited, *Sydney*
Prentice-Hall Canada Inc., *Toronto*
Prentice-Hall Hispanoamericana, S.A., *Mexico*
Prentice-Hall of India Private Limited, *New Delhi*
Prentice-Hall of Japan, Inc., *Tokyo*
Simon & Schuster Asia Pte. Ltd., *Singapore*
Editora Prentice-Hall do Brasil, Ltda., *Rio de Janeiro*

Contents

Problem Set 1.1

Problem Set 1.1

1. Tens

3. Ones

5. Hundreds

7. Hundreds

9. Ones

11. Millions

13. 7 hundreds + 8 tens + 2 ones

15. 9 thousands + 8 hundreds + 3 tens + 4 ones

17. 4 ten thousands + 3 thousands + 8 hundreds + 4 tens + 5 ones

19. 8 thousands + 3 hundreds + 4 ones

21. 8 thousands + 4 hundreds

23. 8 thousands + 9 tens

25. Seventy-five

27. Six hundred eighty-four

29. Seven thousand, eight hundred four

31. Four hundred thousand, seven hundred ninety-six

33. Fifty-one thousand, five

35. Three million, two hundred five thousand, four

37. Seven hundred thousand, thirty-four

39. Ninety thousand, thirty-eight

41. Eight million, two hundred three thousand, four

43. Nine thousand, six

45. Five million, seven hundred ninety-one thousand, two hundred eighty-five

47. 519

49. 4607

51. 704,215

53. 10,003,260

55. 20,008

57. 9020

59. 3,899,515

61. 5,000,004

63. 2,260,000

65. Natural

67. Whole

69. 503,050,010

Problem Set 1.2

1. $$\begin{array}{r} 125 \\ +\underline{713} \\ 838 \end{array}$$

3. $$\begin{array}{r} 4185 \\ +\underline{2113} \\ 6298 \end{array}$$

5. $$\begin{array}{r} 3618 \\ +\underline{\ 221} \\ 3839 \end{array}$$

7. $$\begin{array}{r} 134 \\ +\underline{5165} \\ 5299 \end{array}$$

9. $$\begin{array}{r} 3 \\ 2 \\ +\underline{\ 5} \\ 10 \end{array}$$

11. $$\begin{array}{r} 3 \\ 5 \\ +\underline{\ 4} \\ 12 \end{array}$$

13. $$\begin{array}{r} 6 \\ 8 \\ +\underline{\ 9} \\ 23 \end{array}$$

Problem Set 1.2

15. $\begin{array}{r} 8 \\ 9 \\ +\underline{8} \\ 25 \end{array}$

17. $\begin{array}{r} 372 \\ +\underline{609} \\ 981 \end{array}$

19. $\begin{array}{r} 317 \\ +\underline{259} \\ 576 \end{array}$

21. $\begin{array}{r} 469 \\ +\underline{1} \\ 470 \end{array}$

23. $\begin{array}{r} 340 \\ +\underline{723} \\ 1063 \end{array}$

25. $\begin{array}{r} 325 \\ +\underline{284} \\ 609 \end{array}$

27. $\begin{array}{r} 321 \\ +\underline{929} \\ 1250 \end{array}$

29. $\begin{array}{r} 652 \\ +\underline{359} \\ 1011 \end{array}$

31. $\begin{array}{r} 3245 \\ +\underline{2816} \\ 6061 \end{array}$

33. $\begin{array}{r} 7{,}728 \\ +\underline{5{,}431} \\ 13{,}159 \end{array}$

35. $\begin{array}{r} 31{,}872 \\ +\underline{59{,}341} \\ 91{,}213 \end{array}$

37. $\begin{array}{r} 95{,}410 \\ +\underline{31{,}780} \\ 127{,}190 \end{array}$

39. $\begin{array}{r} 55{,}108 \\ +\underline{1{,}912} \\ 57{,}020 \end{array}$

41. $\begin{array}{r} 321 \\ +\underline{103{,}709} \\ 104{,}030 \end{array}$

43. $\begin{array}{r} 23 \\ 74 \\ +\underline{45} \\ 142 \end{array}$

45. $\begin{array}{r} 38 \\ 24 \\ +\underline{52} \\ 114 \end{array}$

47. $\begin{array}{r} 27 \\ 2 \\ 12 \\ +\underline{38} \\ 79 \end{array}$

49. $\begin{array}{r} 3 \\ 55 \\ 72 \\ +\underline{8} \\ 138 \end{array}$

51. $\begin{array}{r} 728 \\ 324 \\ +\underline{510} \\ 1562 \end{array}$

53. $\begin{array}{r} 4{,}034 \\ 5{,}718 \\ +\underline{2{,}005} \\ 11{,}757 \end{array}$

55. $\begin{array}{r} 3{,}019 \\ 217 \\ 8{,}024 \\ +\underline{\phantom{0{,}0}42} \\ 11{,}302 \end{array}$

57. $\begin{array}{r} 3{,}819{,}032 \\ 405 \\ 2{,}134{,}510 \\ +\underline{\phantom{0{,}000{,}0}29} \\ 5{,}953{,}976 \end{array}$

59. $\begin{array}{r} 908 \\ 7037 \\ +\underline{59} \\ 8004 \end{array}$

61. $\begin{array}{r} 548 \\ 3204 \\ 8 \\ +\underline{59} \\ 3819 \end{array}$

63. $\begin{array}{r} 176 \\ 85 \\ +\underline{189} \\ 450 \end{array}$

Carlos worked 450 hours.

Problem Set 1.2

65.
```
   745
   213
 +1054
  2012
```
Account has $2012.

67.
```
  2314
 +3829
  6143
```
Total attendance was 6143.

69.
```
  184
  316
   54
 +128
  682
```
They purchased 682 pounds of meat.

71.
```
  45,800
 +44,700
  90,500
```
Their combined income is $90,500.

73.
```
  24,322
     324
     278
 +   521
  25,445
```
The odometer reading was 25,445 miles.

75. Sum

77. Add

79. Associative Law

81.
```
   5
   7
   5
 + 3
  20
```

83.
```
   22
   36
   78
 + 64
  200
```

Problem Set 1.3

1.
```
12 - 4
12 = 4 + ?
12 = 4 + 8
12 - 4 = 8
```

3.
```
13 - 5
13 = 5 + ?
13 = 5 + 8
13 - 5 = 8
```

5.
```
16 - 8
16 = 8 + ?
16 = 8 + 8
16 - 8 = 8
```

7.
```
14 - 5
14 = 5 + ?
14 = 5 + 9
14 - 5 = 9
```

9.
```
  84
 -23
  61
```

11.
```
  732
 -231
  501
```

13.
```
  654
 -231
  423
```

15.
```
  6827
 -3614
  3213
```

17. (6 and 5 crossed out)
```
  5 15
  6 5
 -3 7
  2 8
```

19. (6 and 6 crossed out)
```
  5 16
  6 6
 -1 9
  4 7
```

21. (7 and 4 crossed out)
```
    6 14
  3 7 4
 -2 5 7
  1 1 7
```

23. (7 and 4 crossed out)
```
  6 14
  7 4 2
 -3 5 1
  3 9 1
```

25. (9 and 2 crossed out)
```
  8 12
  9 2 4
 -3 7 1
  5 5 3
```

Problem Set 1.3

27.
```
   14
  5 4 12
  6 5 2
 -3 7 4
  2 7 8
```

29.
```
   2 14
  2 3 4 1
 -1 2 5 1
  1 0 9 0
```

31.
```
     13
   8 3 10
  5 9 4 0
 -3 8 7 1
  2 0 6 9
```

33.
```
          12
    4     2 17
  1 5 , 3 7 9
 -  2 , 8 8 0
  1 2 , 4 9 9
```

35.
```
  4 17   7 12
  5 7 , 8 2 9
 -1 8 , 7 4 4
  3 9 , 0 8 5
```

37.
```
  6 10
  7 0
 -3 2
  3 8
```

39.
```
  5 10
  6 0 5
 -3 8 2
  2 2 3
```

41.
```
    6 10
  3 7 0
 -3 2 9
    4 1
```

43.
```
    9
  7 10 12
  8 0 2
 -3 7 4
  4 2 8
```

45.
```
  4 10 7 12
  5 0 8 2
 -4 9 3 6
    1 4 6
```

47.
```
    9 13
  2 10 3 10
  3 0 4 0
 -2 1 5 4
    8 8 6
```

49.
```
    9  9
  7 10 10 10
  8 0 0 0
 -5 1 8 3
  2 8 1 7
```

51.
```
    9  9
  4 10 10 10
  5 0 0 0
 -4 1 3 7
    8 6 3
```

53.
```
    4   10 4 10
  1 5 , 0 5 0
 -1 0 , 4 3 2
    4 , 6 1 8
```

55.
```
    9    9  9
  4 10   10 10 12
  5 0 , 0 0 2
 -2 3 , 7 1 8
  2 6 , 2 8 4
```

57.
```
  6 12
  7 2
 -3 4
  3 8
```
Greg has 38 tapes left.

59.
```
    9
  4 10 10
  5 0 0
 -2 6 7
  2 3 3
```
Shujen needs $233 more.

61.
```
    12
  5 2 14
  6 3 4
 -1 7 9
  4 5 5
```
Cindy has $455 left.

63.
```
      6 15
  1 5 7 5
 -1 2 4 8
    3 2 7
```
Lydia received 327 more paintings in March than in February.

65.
```
        11 9
    3   1 10 11
  2 4 , 2 0 1
 -1 0 , 3 0 4
  1 3 , 8 9 7
```
The odometer reading on the Oldsmobile is 13,897 greater than that on the Buick.

Problem Set 1.3

67. Deductions

$$\begin{array}{r} 327 \\ 143 \\ +\underline{18} \\ 488 \end{array}$$

Net pay = Total pay - Deductions

$$\begin{array}{r} \ 9\ 13 \\ 1\ 1\!\!\!/0\ 3\!\!\!/\ 10 \\ 2\!\!\!/\ 0\!\!\!/\ 4\!\!\!/\ 0\!\!\!/ \\ -\underline{\ 4\ 8\ 8} \\ 1\ 5\ 5\ 2 \end{array}$$

The teacher's net pay is $1552 per month.

69. Difference

71.

$$\begin{array}{r} 13\ \ 16\ 17 \\ 7\ 3\!\!\!/\ \ \ 6\!\!\!/\ 7\!\!\!/\ 17 \\ 8\!\!\!/\ 4\!\!\!/\ ,\ 7\!\!\!/\ 8\!\!\!/\ 7\!\!\!/ \\ -\underline{7\ 8\ ,\ 8\ 9\ 8} \\ 5\ ,\ 8\ 8\ 9 \end{array}$$

73.

$$\begin{array}{r} 9\ \ \ 9\ \ 9 \\ 7\ 1\!\!\!/0\ \ 1\!\!\!/0\ 1\!\!\!/0\ 10 \\ 8\!\!\!/\ 0\!\!\!/\ ,\ 0\!\!\!/\ 0\!\!\!/\ 0\!\!\!/ \\ -\underline{2\ 9\ ,\ 3\ 5\ 2} \\ 5\ 0\ ,\ 6\ 4\ 8 \end{array}$$

Problem Set 1.4

1. $37 \approx 40$

3. $72 \approx 70$

5. $625 \approx 630$

7. $678 \approx 680$

9. $732 \approx 700$

11. $675 \approx 700$

13. $85 \approx 100$

15. $6792 \approx 6800$

17. $3821 \approx 4000$

19. $6342 \approx 6000$

21. $784 \approx 1000$

23. $17,520 \approx 18,000$

25.

$$\begin{array}{rcr} 52 & \approx & 50 \\ 35 & \approx & 40 \\ 74 & \approx & 70 \\ +\underline{82} & \approx & +\underline{80} \\ & & 240 \end{array}$$

27.

$$\begin{array}{rcr} 571 & \approx & 570 \\ 84 & \approx & 80 \\ 219 & \approx & 220 \\ +\underline{32} & \approx & +\underline{30} \\ & & 900 \end{array}$$

29.

$$\begin{array}{rcr} 725 & \approx & 730 \\ -\underline{386} & \approx & -\underline{390} \\ & & 340 \end{array}$$

31.

$$\begin{array}{rcr} 7834 & \approx & 7830 \\ -\underline{5177} & \approx & -\underline{5180} \\ & & 2650 \end{array}$$

33.

$$\begin{array}{rcr} 378 & \approx & 400 \\ 213 & \approx & 200 \\ 655 & \approx & 700 \\ +\underline{472} & \approx & +\underline{500} \\ & & 1800 \end{array}$$

35.

$$\begin{array}{rcr} 729 & \approx & 700 \\ 85 & \approx & 100 \\ 612 & \approx & 600 \\ +\underline{92} & \approx & +\underline{100} \\ & & 1500 \end{array}$$

37.

$$\begin{array}{rcr} 1312 & \approx & 1300 \\ 759 & \approx & 800 \\ 62 & \approx & 100 \\ +\underline{820} & \approx & +\underline{800} \\ & & 3000 \end{array}$$

39.

$$\begin{array}{rcr} 6182 & \approx & 6200 \\ -\underline{3241} & \approx & -\underline{3200} \\ & & 3000 \end{array}$$

41.

$$\begin{array}{rcr} 5,784 & \approx & 6,000 \\ 2,346 & \approx & 2,000 \\ 89,340 & \approx & 89,000 \\ +\underline{765} & \approx & +\underline{1,000} \\ & & 98,000 \end{array}$$

Problem Set 1.4

43.
$$\begin{array}{rcr} 15{,}940 & \approx & 16{,}000 \\ -\underline{3{,}687} & \approx & -\underline{4{,}000} \\ & & 12{,}000 \end{array}$$

45. $7358 \approx 7400$
The approximate taxes are $7400.

47. $20{,}320 \approx 20{,}000$
The approximate height is 20,000 feet.

49.
$$\begin{array}{rcr} 8 & \approx & 10 \\ 74 & \approx & 70 \\ +\underline{105} & \approx & +\underline{110} \\ & & 190 \end{array}$$
The estimate of the total amount is $190.

51.
$$\begin{array}{rcr} 875 & \approx & 900 \\ 435 & \approx & 400 \\ +\underline{365} & \approx & +\underline{400} \\ & & 1700 \end{array}$$
The approximate total is $1700.

53. $5 < 11$

55. $27 > 18$

57. $5 > 0$

59. $112 < 126$

61. $81 > 77$

63. $1214 > 1102$

65. Less

67. False

69. True

Problem Set 1.5

1.
$$\begin{array}{r} 52 \\ \times\underline{3} \\ 156 \end{array}$$

3.
$$\begin{array}{r} 43 \\ \times\underline{3} \\ 129 \end{array}$$

5.
$$\begin{array}{r} 74 \\ \times\underline{3} \\ 222 \end{array}$$

7.
$$\begin{array}{r} 56 \\ \times\underline{5} \\ 280 \end{array}$$

9.
$$\begin{array}{r} 515 \\ \times\underline{4} \\ 2060 \end{array}$$

11.
$$\begin{array}{r} 7234 \\ \times\underline{4} \\ 28{,}936 \end{array}$$

13.
$$\begin{array}{r} 37 \\ \times\underline{28} \\ 296 \\ \underline{74} \\ 1036 \end{array}$$

15.
$$\begin{array}{r} 87 \\ \times\underline{32} \\ 174 \\ \underline{261} \\ 2784 \end{array}$$

17.
$$\begin{array}{r} 324 \\ \times\underline{49} \\ 2916 \\ \underline{1296} \\ 15876 \end{array}$$

19.
$$\begin{array}{r} 455 \\ \times\underline{54} \\ 1820 \\ \underline{2275} \\ 24570 \end{array}$$

21.
$$\begin{array}{r} 519 \\ \times\underline{236} \\ 3114 \\ 1557 \\ \underline{1038} \\ 122484 \end{array}$$

23.
$$\begin{array}{r} 695 \\ \times\underline{438} \\ 5560 \\ 2085 \\ \underline{2780} \\ 304410 \end{array}$$

25.
$$\begin{array}{r} 8945 \\ \times\underline{235} \\ 44725 \\ 26835 \\ \underline{17890} \\ 2102075 \end{array}$$

Problem Set 1.5

27.
  6932
× 2314
 27728
 6932
20796
13864
16040648

29.
  72
× 30
2160

31.
  517
× 305
2585
1551
157685

33.
  679
× 402
1358
2716
272958

35.
  272
× 480
21760
1088
130560

37.
  575
× 240
23000
1150
138000

39.
  5148
× 4700
3603600
20592
24195600

41.
  520
× 37
3640
1560
19240

43.
  840
× 50
42000

45.
  802
× 504
3208
4010
404208

47.
  3070
× 402
6140
12280
1234140

49. $(4 \times 3) \times 2 = 12 \times 2$
$= 24$
$4 \times (3 \times 2) = 4 \times 6$
$= 24$

51. $(6 \times 3) \times 8 = 18 \times 8$
$= 144$
$6 \times (3 \times 8) = 6 \times 24$
$= 144$

53. $(7 \times 9) \times 3 = 63 \times 3$
$= 189$
$7 \times (9 \times 3) = 7 \times 27$
$= 189$

55. $(6 \times 8) \times 2 = 48 \times 2$
$= 96$
$6 \times (8 \times 2) = 6 \times 16$
$= 96$

57. $9 \times 18 = 162$
Jeff spent $162 on shirts.

59. $85 \times 67 = 5695$
The area is 5695 square centimeters.

61. $348 \times 14 = 4872$
Sharon earned $4872.

63. $568 \times 12 = 6816$
Sean spends $6816 on rent.

65. $279 \times 365 = 101,835$
Blue Linen Company made 101,835 bath towels.

67. Multiplied

69. Commutative

71. Length, width

73. $8 \times 12 = 96$
Manuel paid $96 for the compact discs.

75. $329 + 329 + 287 + 287 = 1232$
The perimeter is 1232 meters.

Problem Set 1.6

1. $42 = 7 \times 6$

Problem Set 1.6

3. $81 = 9 \times 9$

5. $54 = 9 \times 6$

7. $64 = 8 \times 8$

9. $42 = 42 \div 1$
   $1 = 42 \div 42$

11. $9 = 27 \div 3$
    $3 = 27 \div 9$

13. $7 = 63 \div 7$
    $9 = 63 \div 9$

15. $18 = 18 \div 1$
    $1 = 18 \div 18$

17.
```
   281
 3)843
   6
   24
   24
    03
     3
     0
```
The answer is 281.

19.
```
   1151
 5)5755
   5
   07
    5
    25
    25
     05
      5
      0
```
The answer is 1151.

21.
```
    249
 6)1498
   12
    29
    24
     58
     54
      4
```
The answer is 249 R 4.

23.
```
    198
 9)1784
   9
   88
   81
    74
    72
     2
```
The answer is 198 R 2.

25.
```
    98
 3)295
   27
    25
    24
     1
```
The answer is 98 R 1.

27.
```
    73
 8)585
   56
    25
    24
     1
```
The answer is 73 R 1.

29.
```
    398
 6)2389
   18
    58
    54
     49
     48
      1
```
The answer is 398 R 1.

31.
```
    961
 4)3845
   36
    24
    24
     05
      4
      1
```
The answer is 961 R 1.

33.
```
      55
 46)2530
    230
     230
     230
       0
```
The answer is 55.

35.
```
      34
 60)2084
    180
     284
     240
      44
```
The answer is 34 R 44.

37.
```
     26
 34)898
    68
    218
    204
     14
```
The answer is 26 R 14.

Problem Set 1.6

39.
```
   109
 4)439
   4
   039
    36
     3
```
The answer is 109 R 3.

41.
```
    350
  6)2105
    18
     30
     30
      05
```
The answer is 350 R 5.

43.
```
      30
 62)1860
    186
      00
```
The answer is 30.

45.
```
     103
 48)4944
    48
     144
     144
       0
```
The answer is 103.

47.
```
      56
 37)2098
    185
     248
     222
      26
```
The answer is 56 R 26.

49.
```
      42
 56)2384
    224
     144
     112
      32
```
The answer is 42 R 32.

51.
```
      206
 72)14895
    144
      495
      432
       63
```
The answer is 206 R 63.

53.
```
      305
 36)10989
    108
      189
      180
        9
```
The answer is 305 R 9.

55.
```
      509
 53)26989
    265
      489
      477
       12
```
The answer is 509 R 12.

57.
```
        721
 321)231441
     2247
       674
       642
        321
        321
          0
```
The answer is 721.

59.
```
        306
 494)151178
     1482
       2978
       2964
         14
```
The answer is 306 R 14.

61.
```
    892
  6)5352
    48
     55
     54
      12
      12
       0
```
Teresa's monthly payment was $892.

63.
```
     3500
 17)59500
    51
     85
     85
      000
```
The machine caps 3500 bottles in 1 hour.

Problem Set 1.6

65.

$$\begin{array}{r} 45 \\ 18\overline{)810} \\ \underline{72} \\ 90 \\ \underline{90} \\ 0 \end{array}$$

Each ticket costs $45.

67. Quotient

69. Dividend

71.

$$\begin{array}{r} 5 \\ 3\overline{)17} \\ \underline{15} \\ 2 \end{array}$$

Monique can make 5 tablecloths; there is a remainder of 2 yards of fabric.

Problem Set 1.7

1.

$$\begin{array}{r} 735 \\ +\underline{\ 879} \\ 1614 \end{array}$$

Linda earned $1614.

3.

$$\begin{array}{r} 224 \\ 121 \\ +\underline{187} \\ 532 \end{array}$$

The total area is 532 square feet.

5.

$$\begin{array}{r} 39,400,000 \\ +\underline{38,700,000} \\ 78,100,000 \end{array}$$

The Highway Airport served 78,100,000 passengers.

7.

$$\begin{array}{r} 125 \\ -\underline{109} \\ 16 \end{array}$$

Rhonda had 16 cookies left.

9.

$$\begin{array}{r} 39,400,000 \\ -\underline{38,700,000} \\ 700,000 \end{array}$$

The difference in the number of passengers served is 700,000.

11. Total - Checks = Balance

Checks

$$\begin{array}{r} 74 \\ 295 \\ +\underline{\ 27} \\ 396 \end{array}$$

Balance

$$\begin{array}{r} 878 \\ -\underline{396} \\ 482 \end{array}$$

Swan has $482 left in her checking account.

13.

$$\begin{array}{r} 12 \\ \times\underline{\ 8} \\ 96 \end{array}$$

Sarah paid $96 for the compact discs.

15.

$$\begin{array}{r} 23 \\ \times\underline{12} \\ 46 \\ \underline{23\ \ } \\ 276 \end{array}$$

There are 276 eggs in 23 dozen eggs.

17.

$$\begin{array}{r} 89 \\ \times\underline{74} \\ 356 \\ \underline{623\ } \\ 6586 \end{array}$$

The area of the rectangle is 6586 square meters.

19.

$$\begin{array}{r} 245 \\ \times\underline{\ \ 3} \\ 735 \end{array} \qquad \begin{array}{r} 370 \\ \times\underline{\ \ 5} \\ 1850 \end{array} \qquad \begin{array}{r} 735 \\ +\underline{1850} \\ 2585 \end{array}$$

Juan paid $2585 for the paintings.

21.

$$\begin{array}{r} 17 \\ 8\overline{)137} \\ \underline{8\ } \\ 57 \\ \underline{56} \\ 1 \end{array}$$

Each boy received 17 cookies; Jim was left with 1 cookie.

23.

$$\begin{array}{r} 3 \\ 35\overline{)105} \\ \underline{105} \\ 0 \end{array}$$

There are three sections of second grade.

Problem Set 1.7

25.
```
     14
16)229
   16
    69
    64
     5
```
Each plate should have 14 cupcakes; there are 5 cupcakes left over.

27.
```
     22
16)352
   32
    32
    32
     0
```
The car gets 22 miles per gallon of gasoline.

29. Balance = Charges - Payment
```
Charges    75   Balance   460
           38            -236
         +347             224
          460
```
Ken owes $224.

31.
```
 115      162
+ 47     ×  6
 162      972
```
Mary pays $972 for food and utilities for 6 months.

33.
```
  35        14
×  8    20)280
 280       20
            80
            80
             0
```
Susan used 14 $20 bills to pay for the dresses.

35.
```
  75       32      2700
× 36     × 69     -2208
 450      288       492
225      192
2700     2208
```
The Lamprey's left over space is 492 square feet.

37.
```
Vinyl      Carpeting
     12            14
16)192        34)476
   16            34
    32           136
    32           136
     0             0
```
The carpeting is $2 per square yard more expensive than the vinyl.

39. Addition

41. Multiplication

43.
```
   243000
 ×     16
  1458000
  243000
  3888000
```
The total value of the homes is $3,888,000.

Problem Set 1.8

1. $16 + (8 - 7) = 16 + 1$
$= 17$

3. $57 - (13 - 4) = 57 - 9$
$= 48$

5. $4 + 32 \div 8 = 4 + 4$
$= 8$

7. $7 \bullet 8 - 5 = 56 - 5$
$= 51$

9. $(76 - 12) - 27 = 64 - 27$
$= 37$

11. $97 - 8 \bullet 9 = 97 - 72$
$= 25$

13. $25 - 8 - 3 = 17 - 3$
$= 14$

15. $54 \div 9 \div 3 = 6 \div 3$
$= 2$

17. $6 \bullet 8 \div 2 = 48 \div 2$
$= 24$

19. $45 \div 5 \bullet 4 = 9 \bullet 4$
$= 36$

21. $(14 + 16) - 3 \bullet 9 = 30 - 3 \bullet 9$
$= 30 - 27$
$= 3$

23. $7 + 8 \bullet 9 = 7 + 72$
$= 79$

25. $28 + [12 - (6 - 3)] = 28 + [12 - 3]$
$= 28 + 9$
$= 37$

27. $6 + [4 \bullet (8 - 5)] = 6 + [4 \bullet 3]$
$= 6 + 12$
$= 18$

Problem Set 1.8

29. $3 + \{10 + [37 - (18 - 4)]\}$
$= 3 + \{10 + [37 - 14]\}$
$= 3 + \{10 + 23\}$
$= 3 + 33$
$= 36$

31. Simplify

33. Brackets

35. Innermost

37. $8 + 4 \bullet 4 \bullet 4 \div (9 - 5)$
$= 8 + 4 \bullet 4 \bullet 4 \div 4$
$= 8 + 16 \bullet 4 \div 4$
$= 8 + 64 \div 4$
$= 8 + 16$
$= 24$

39. $3 + [26 - (9 + 6) \div 5] - (82 - 7 \bullet 9)$
$= 3 + [26 - 15 \div 5] - (82 - 7 \bullet 9)$
$= 3 + [26 - 3] - (82 - 63)$
$= 3 + 23 - 19$
$= 7$

Chapter 1 Additional Exercises

1. Tens

3. Thousands

5. 6 ten thousands + 7 thousands + 8 hundreds + 2 ones

7. Seven hundred four thousand, eight hundred twenty-nine

9. 5060

11.
```
 3782
+ 213
 3995
```

13.
```
  329
+ 881
 1210
```

15.
```
  723
 3691
+ 428
 4842
```

17.
```
  65,307
     208
 729,823
+     42
 795,380
```

19.
```
  57
  24
  38
+ 29
 148
```
Ben spent $148.

21.
```
 738
-614
 124
```

23.
```
     11
   8 1̸ 18
 5 9̸ 2̸ 8̸
-3 4 9 9
 2 4 2 9
```

25.
```
   9  9
 7 10 10 10
 8̸ 0̸ 0̸ 0̸
-6 1 8 9
 1 8 1 1
```

27.
```
   9 13   9  9
 0 10 3̸  10 10 10
 1̸ 0̸ 4̸ , 0̸ 0̸ 0̸
-  9 6 , 0 5 7
     7 , 9 4 3
```

29.
```
   12 12
 6 2̸ 2̸ 12
 7̸ 3̸ 3̸ 2̸
-5 7 8 4
 1 5 4 8
```
There are 1548 more night students than day students.

31.
```
  485 ≈   490
   32 ≈    30
   56 ≈    60
+ 519 ≈ + 520
         1100
```

33.
```
 7831 ≈  7800
- 854 ≈ - 900
         6900
```

35.
```
 18,948 ≈  19,000
  3,579 ≈   4,000
+ 2,483 ≈ + 2,000
           25,000
```

37. $19 > 8$

39. $923 < 974$

Chapter 1 Additional Exercises

41.
$2379 \approx 2400$
$307 \approx 300$
$+1528 \approx +1500$
4200
Jeff flew approximately 4200 miles.

43.
623
× 5
3115

45.
78
× 89
702
624
6942

47.
272
× 380
21760
816
103360

49.
313
×302
626
939
94526

51.
$5 \times 9 \times 2 = 45 \times 2$
$= 90$

53.
77
× 68
616
462
5236
The area is 5236 square feet.

55.
43
8)347
32
27
24
3
The answer is 43 R 3.

57.
809
9)7285
72
085
81
4
The answer is 809 R 4.

59.
1668
32)53402
32
214
192
220
192
282
256
26
The answer is 1668 R 26.

61.
2106
236)497031
472
250
236
1431
1416
15
The answer is 2106 R 15.

63.
85
97
92
+ 78
352

88
4)352
32
32
32
0
Janelle's average score is 88.

65.
29
× 7
203
There are 203 days in 29 weeks.

67.
17
25)425
25
175
175
0
Jeanne's car takes 17 gallons to travel 425 miles.

69.
517
× 3
1551

482
× 4
1928

1551
+1928
3479
Gil drove 3479 miles.

71.
$3 + 7 \bullet 8 = 3 + 56$
$= 59$

73.
$64 \div 8 \bullet 4 = 8 \bullet 4$
$= 32$

75.
$27 - [14 - 2 \bullet 3] = 27 - [14 - 6]$
$= 27 - 8$
$= 19$

Chapter 1 Additional Exercises

77. $4 + \{12 - [9 - (2+3)]\}$
$= 4 + \{12 - [9 - 5]\}$
$= 4 + \{12 - 4\}$
$= 4 + 8$
$= 12$

Chapter 1 Practice Test

1. Thousands

3.
$$\begin{array}{r} 782 \\ +\underline{216} \\ 998 \end{array}$$

5.
$$\begin{array}{r} 327 \\ 24 \\ +\underline{4628} \\ 4979 \end{array}$$

7.
$$\begin{array}{r} 7\ 15\ 1\ 14 \\ \not{8}\ \not{5}\ \not{2}\ \not{4} \\ -\underline{3\ 7\ 1\ 8} \\ 4\ 8\ 0\ 6 \end{array}$$

9.
$$\begin{array}{rcr} 785 & \approx & 800 \\ +\underline{3642} & \approx & +\underline{3600} \\ & & 4400 \end{array}$$

11.
$$\begin{array}{r} 432 \\ \times\underline{\quad 7} \\ 3024 \end{array}$$

13.
$$\begin{array}{r} 627 \\ \times\underline{\quad 300} \\ 188100 \end{array}$$

15.
$$\begin{array}{r} 79 \\ 9\overline{)713} \\ \underline{63} \\ 83 \\ \underline{81} \\ 2 \end{array}$$
The answer is 79 R 2.

17.
$$\begin{array}{r} 74 \\ \times\underline{\ 60} \\ 4440 \end{array}$$
There are 4440 seconds in 74 minutes.

19.
$$\begin{array}{r} 56 \\ 34\overline{)1904} \\ \underline{170} \\ 204 \\ \underline{204} \\ 0 \end{array}$$
Teresa's car takes 56 gallons to go 1904 miles.

21. $8 \bullet 9 \div 3 = 72 \div 3$
$= 24$

Problem Set 2.1

1. $\frac{3}{4}$

3. $\frac{5}{4}$

5. $\frac{1}{12}$

7. $\frac{6}{5}$

9. numerator = 3
denominator = 5

11. numerator = 23
denominator = 12

13. numerator = 7
denominator = 11

15. proper fractions
$= \frac{7}{8}, \frac{3}{4}, \frac{19}{21}$
improper fractions $= \frac{18}{17}, \frac{6}{6}$

17. proper fractions
$= \frac{5}{9}, \frac{1}{7}, \frac{12}{35}, \frac{2}{9}$
improper fractions $= \frac{11}{11}$

19. $\frac{3}{3} = 1$

21. $\frac{9}{1} = 9$

23. $\frac{0}{14} = 0$

25. $\frac{39}{39} = 1$

27. $\frac{586}{1} = 586$

29. $\frac{0}{254} = 0$

31. eight, fifteen

Problem Set 2.1

33. numerator

35. bar

37. improper

39. $\frac{22}{37}$

41.
$$\begin{array}{r} 89 \\ \times\underline{24} \\ 356 \\ \underline{178} \\ 2136 \end{array}$$

43.
$$\begin{array}{r} 629 \\ \times\underline{587} \\ 4403 \\ 5032 \\ \underline{3145} \\ 369223 \end{array}$$

45.
$$\begin{array}{r} 890 \\ \times\underline{506} \\ 5340 \\ \underline{4450} \\ 450340 \end{array}$$

47. $7 \times 9 \times 2 = 63 \times 2$
$= 126$

49. zero

Problem Set 2.2

1.
$$\begin{array}{r} 16 \\ 2\overline{)32} \\ \underline{2} \\ 12 \\ \underline{12} \\ 0 \end{array}$$
32 is divisible by 2 since there is no remainder.

3.
$$\begin{array}{r} 3 \\ 5\overline{)16} \\ \underline{15} \\ 1 \end{array}$$
16 is not divisible by 5 since there is a remainder.

5.
$$\begin{array}{r} 4 \\ 13\overline{)52} \\ \underline{52} \\ 0 \end{array}$$
52 is divisible by 13 since there is no remainder.

7. 2 is prime

9. 4 = 2 • 2; 4 is composite

11. 3 is prime

13. 58 = 2 • 29; 58 is composite

15. 11 is prime

17. 13 is prime

19. 6 = 2 • 3

21. 8 = 2 • 2 • 2

23. 21 = 3 • 7

25. 35 = 5 • 7

27. 120 = 2 • 2 • 2 • 3 • 5

29. 100 = 2 • 2 • 5 • 5

31. 77 = 7 • 11

33. 154 = 2 • 7 • 11

35. 90 = 2 • 3 • 3 • 5

37. 65 = 5 • 13

39. 169 = 13 • 13

41. 125 = 5 • 5 • 5

43. remainder

45. prime

47. 161 = 7 • 23

49.
$$\begin{array}{r} 307 \\ 3\overline{)921} \\ \underline{9} \\ 021 \\ \underline{21} \\ 0 \end{array}$$
The answer is 307.

51.
$$\begin{array}{r} 1759 \\ 5\overline{)8795} \\ \underline{5} \\ 37 \\ \underline{35} \\ 29 \\ \underline{25} \\ 45 \\ \underline{45} \\ 0 \end{array}$$
The answer is 1759.

Problem Set 2.2

53.
```
     263
32)8423
   64
   202
   192
    103
     96
      7
```
The answer is 263 R 7.

55.
```
     12
65)784
   65
   134
   130
     4
```
The answer is 12 R 4.

57.
```
     273
25)6832
   50
   183
   175
    82
    75
     7
```
The answer is 273 R 7.

21.
```
   534
×   27
  3738
 1068
 14418
```

23.
```
   625
×  503
  1875
3125
314375
```

25. $4 \times 8 \times 9 = 32 \times 9$
$= 288$

27.
```
  18
×  6
 108
```
Carrie paid $108 for the scarves.

29.
```
  84     975
×  8    -672
 672     303
```
Jim has $303 left to spend.

Problem Set 2.3

1. 338, 84, 52, 860

3. 3405, 860, 7185

5. 33, 6594, 16,785, 594

7. 470, 40

9. 36, 260, 428, 512,818

11. 260, 625, 3915

13. two

15. five

17. 27, 180, and 7281 are divisible by 9.
27; 2 + 7 = 9
180; 1 + 8 + 0 = 9
7281; 7 + 2 + 8 + 1 = 18
A number is divisible by nine if the sum of its digits is divisible by nine.

19. A number is divisible by 8 if the number named by the last 3 digits is divisible by 8.

Problem Set 2.4

1. $\frac{3}{8} \times \frac{1}{5} = \frac{3 \times 1}{8 \times 5} = \frac{3}{40}$

3. $\frac{5}{6} \times \frac{7}{9} = \frac{5 \times 7}{6 \times 9} = \frac{35}{54}$

5. $\frac{7}{4} \bullet \frac{3}{2} = \frac{7 \bullet 3}{4 \bullet 2} = \frac{21}{8}$

7. $\frac{11}{8} \bullet \frac{4}{3} = \frac{11 \bullet 4}{8 \bullet 3} = \frac{44}{24}$

9. $\frac{1}{4} \bullet \frac{9}{8} = \frac{1 \bullet 9}{4 \bullet 8} = \frac{9}{32}$

11. $\frac{8}{7} \bullet \frac{1}{3} = \frac{8 \bullet 1}{7 \bullet 3} = \frac{8}{21}$

13. $8 \bullet \frac{5}{2} = \frac{8}{1} \bullet \frac{5}{2} = \frac{8 \bullet 5}{1 \bullet 2} = \frac{40}{2}$

15. $9 \bullet \frac{7}{8} = \frac{9}{1} \bullet \frac{7}{8} = \frac{9 \bullet 7}{1 \bullet 8} = \frac{63}{8}$

17. $\frac{1}{8} \bullet 3 = \frac{1}{8} \bullet \frac{3}{1} = \frac{1 \bullet 3}{8 \bullet 1} = \frac{3}{8}$

19. $\frac{3}{4} \bullet 7 = \frac{3}{4} \bullet \frac{7}{1} = \frac{3 \bullet 7}{4 \bullet 1} = \frac{21}{4}$

Problem Set 2.4

21. $\frac{1}{5} \bullet \frac{1}{4} = \frac{1 \bullet 1}{5 \bullet 4} = \frac{1}{20}$

23. $2 \bullet \frac{5}{3} = \frac{2}{1} \bullet \frac{5}{3} = \frac{2 \bullet 5}{1 \bullet 3} = \frac{10}{3}$

25. $3 \bullet \frac{11}{4} = \frac{3}{1} \bullet \frac{11}{4} = \frac{3 \bullet 11}{1 \bullet 4} = \frac{33}{4}$

27. $\frac{9}{15} \bullet \frac{3}{4} = \frac{9 \bullet 3}{15 \bullet 4} = \frac{27}{60}$

29. $\frac{5}{4} \bullet \frac{5}{2} = \frac{5 \bullet 5}{4 \bullet 2} = \frac{25}{8}$
The area is $\frac{25}{8}$ square feet.

31. $2 \bullet \frac{2}{3} = \frac{2}{1} \bullet \frac{2}{3} = \frac{2 \bullet 2}{1 \bullet 3} = \frac{4}{3}$
Jane should use $\frac{4}{3}$ cup of tomato paste.

33. numerators, denominators

35. multiplication

37. $\frac{46}{3} \times \frac{6}{59} = \frac{46 \times 6}{3 \times 59} = \frac{276}{177}$

39. $104 \bullet \frac{5}{3} = \frac{104}{1} \bullet \frac{5}{3}$
$= \frac{104 \bullet 5}{1 \bullet 3} = \frac{520}{3}$

41. 274 + 82 + 406 = 762
974 - 762 = 212
Dale has $212 left in his checking account.

43. 375 + 46 = 421
421 × 8 = 3368
Hoang pays $3368 for rent and parking for 8 months.

45.
```
 64,200
-50,750
 13,450
```
The difference between their two salaries is $13,450.

47.
```
      5
74)372
   370
     2
```
There will be 5 puppets in each box; there will be 2 puppets left over.

49.
```
  17          10
 × 6     10)102
 102         10
             02
```
Zelda used eleven $10 bills to pay for the scarves; she received $8 in change.

Problem Set 2.5

1. $\frac{4}{8} = \frac{1 \bullet \cancel{4}}{2 \bullet \cancel{4}} = \frac{1}{2}$

3. $\frac{6}{9} = \frac{2 \bullet \cancel{3}}{3 \bullet \cancel{3}} = \frac{2}{3}$

5. $\frac{9}{12} = \frac{\cancel{3} \bullet 3}{\cancel{3} \bullet 4} = \frac{3}{4}$

7. $\frac{12}{16} = \frac{3 \bullet \cancel{4}}{4 \bullet \cancel{4}} = \frac{3}{4}$

9. $\frac{35}{42} = \frac{5 \bullet \cancel{7}}{6 \bullet \cancel{7}} = \frac{5}{6}$

11. $\frac{90}{150} = \frac{3 \bullet \cancel{3} \bullet \cancel{10}}{\cancel{3} \bullet 5 \bullet \cancel{10}} = \frac{3}{5}$

13. $\frac{11}{22} = \frac{1 \bullet \cancel{11}}{2 \bullet \cancel{11}} = \frac{1}{2}$

15. $\frac{63}{72} = \frac{7 \bullet \cancel{9}}{8 \bullet \cancel{9}} = \frac{7}{8}$

17. $\frac{75}{125} = \frac{3 \bullet \cancel{25}}{5 \bullet \cancel{25}} = \frac{3}{5}$

19. $\frac{130}{260} = \frac{1 \bullet \cancel{130}}{2 \bullet \cancel{130}} = \frac{1}{2}$

21. Yes, since $2 \bullet 12 = 3 \bullet 8$
$24 = 24$

23. No, since $3 \bullet 4 \neq 16 \bullet 1$
$12 \neq 16$

25. Yes, since $5 \bullet 16 = 8 \bullet 10$
$80 = 80$

27. No, since $20 \bullet 2 \neq 3 \bullet 5$
$40 \neq 15$

29. Yes, since $7 \bullet 48 = 8 \bullet 42$
$336 = 336$

31. Yes, since $36 \bullet 4 = 48 \bullet 3$
$144 = 144$

Problem Set 2.5

33. $3 \cdot \frac{1}{3} = \frac{3}{1} \cdot \frac{1}{3} = \frac{\cancel{3} \cdot 1}{1 \cdot \cancel{3}} = 1$

35. $\frac{5}{8} \cdot \frac{1}{5} = \frac{\cancel{5} \cdot 1}{8 \cdot \cancel{5}} = \frac{1}{8}$

37. $\frac{1}{4} \cdot \frac{2}{3} = \frac{1 \cdot 2}{4 \cdot 3} = \frac{1 \cdot \cancel{2}}{\cancel{2} \cdot 2 \cdot 3} = \frac{1}{6}$

39. $\frac{3}{4} \cdot \frac{1}{9} = \frac{3 \cdot 1}{4 \cdot 9}$
$= \frac{\cancel{3} \cdot 1}{4 \cdot 3 \cdot \cancel{3}} = \frac{1}{12}$

41. $\frac{7}{8} \cdot \frac{4}{5} = \frac{7 \cdot 4}{8 \cdot 5}$
$= \frac{7 \cdot \cancel{4}}{2 \cdot \cancel{4} \cdot 5} = \frac{7}{10}$

43. $\frac{5}{9} \cdot \frac{18}{21} = \frac{5 \cdot 18}{9 \cdot 21}$
$= \frac{5 \cdot 2 \cdot \cancel{9}}{\cancel{9} \cdot 21} = \frac{10}{21}$

45. $\frac{6}{15} \cdot \frac{5}{22} = \frac{6 \cdot 5}{15 \cdot 22}$
$= \frac{\cancel{2} \cdot \cancel{3} \cdot \cancel{5} \cdot 1}{\cancel{3} \cdot \cancel{5} \cdot \cancel{2} \cdot 11} = \frac{1}{11}$

47. $\frac{9}{4} \cdot \frac{8}{6} = \frac{9 \cdot 8}{4 \cdot 6}$
$= \frac{3 \cdot \cancel{3} \cdot \cancel{2} \cdot \cancel{4}}{\cancel{4} \cdot \cancel{3} \cdot \cancel{2} \cdot 1} = \frac{3}{1} = 3$

49. $8 \cdot \frac{1}{9} = \frac{8}{1} \cdot \frac{1}{9} = \frac{8 \cdot 1}{1 \cdot 9} = \frac{8}{9}$

51. $9 \cdot \frac{1}{9} = \frac{9}{1} \cdot \frac{1}{9} = \frac{\cancel{9} \cdot 1}{1 \cdot \cancel{9}} = 1$

53. $\frac{5}{8} \cdot \frac{8}{5} = \frac{\cancel{5} \cdot \cancel{8} \cdot 1}{\cancel{8} \cdot \cancel{5} \cdot 1} = 1$

55. $\frac{10}{3} \cdot \frac{3}{10} = \frac{\cancel{10} \cdot \cancel{3} \cdot 1}{\cancel{3} \cdot \cancel{10} \cdot 1} = 1$

57. $\frac{1}{3} \cdot 3 = \frac{1}{3} \cdot \frac{3}{1} = \frac{1 \cdot \cancel{3}}{\cancel{3} \cdot 1} = 1$

59. $\frac{8}{3} \cdot \frac{3}{8} = \frac{\cancel{8} \cdot \cancel{3} \cdot 1}{\cancel{3} \cdot \cancel{8} \cdot 1} = 1$

61. $\frac{1}{8} \cdot 8 = \frac{1}{8} \cdot \frac{8}{1} = \frac{1 \cdot \cancel{8}}{\cancel{8} \cdot 1} = 1$

63. $\frac{2}{3} \cdot 15 = \frac{2}{3} \cdot \frac{15}{1} = \frac{2 \cdot 15}{3 \cdot 1}$
$= \frac{2 \cdot \cancel{3} \cdot 5}{\cancel{3} \cdot 1} = \frac{10}{1} = 10$

65. $\frac{7}{8} \cdot 32 = \frac{7}{8} \cdot \frac{32}{1} = \frac{7 \cdot 32}{8 \cdot 1}$
$= \frac{7 \cdot 4 \cdot \cancel{8}}{\cancel{8} \cdot 1} = \frac{28}{1} = 28$

67. $\frac{7}{8} \cdot \frac{5}{21} = \frac{7 \cdot 5}{8 \cdot 21}$
$= \frac{\cancel{7} \cdot 5}{8 \cdot 3 \cdot \cancel{7}} = \frac{5}{24}$

69. $\frac{3}{4} \cdot \frac{2}{45} = \frac{3 \cdot 2}{4 \cdot 45}$
$= \frac{\cancel{3} \cdot \cancel{2} \cdot 1}{\cancel{2} \cdot 2 \cdot \cancel{3} \cdot 15} = \frac{1}{30}$

71. $\frac{1}{5} \cdot 480 = \frac{1}{5} \cdot \frac{480}{1} = \frac{1 \cdot 480}{5 \cdot 1}$
$= \frac{1 \cdot \cancel{5} \cdot 96}{\cancel{5} \cdot 1} = \frac{96}{1} = 96$

73. $\frac{1}{5} \cdot 4000 = \frac{1}{5} \cdot \frac{4000}{1}$
$= \frac{1 \cdot 4000}{5 \cdot 1} = \frac{1 \cdot \cancel{5} \cdot 800}{\cancel{5} \cdot 1}$
$= \frac{800}{1} = 800$

75. $\frac{2}{3} \cdot 2700 = \frac{2}{3} \cdot \frac{2700}{1}$
$= \frac{2 \cdot 2700}{3 \cdot 1} = \frac{2 \cdot \cancel{3} \cdot 900}{\cancel{3} \cdot 1}$
$= \frac{1800}{1} = 1800$

77. $8 \cdot \frac{1}{4} = \frac{8}{1} \cdot \frac{1}{4} = \frac{8 \cdot 1}{1 \cdot 4}$
$= \frac{2 \cdot \cancel{4} \cdot 1}{1 \cdot \cancel{4}} = \frac{2}{1} = 2$

The $\frac{1}{4}$ pound of steak costs $2.

79. $\frac{3}{4} \cdot \frac{3}{8} = \frac{3 \cdot 3}{4 \cdot 8} = \frac{9}{32}$

Karen used $\frac{9}{32}$ gallon of milk.

Problem Set 2.5

81. $32 \cdot \frac{7}{8} = \frac{32}{1} \cdot \frac{7}{8} = \frac{32 \cdot 7}{1 \cdot 8}$
$= \frac{4 \cdot \cancel{8} \cdot 7}{1 \cdot \cancel{8}} = \frac{28}{1} = 28$
Jeanine's car will travel 28 miles on $\frac{7}{8}$ gallon of gas.

83. $400 \cdot \frac{3}{10} = \frac{400}{1} \cdot \frac{3}{10}$
$= \frac{400 \cdot 3}{1 \cdot 10} = \frac{40 \cdot \cancel{10} \cdot 3}{1 \cdot \cancel{10}}$
$= \frac{120}{1} = 120$
Erika budgeted \$120 per week for food.

85. $14{,}000 \cdot \frac{2}{7} = \frac{14{,}000}{1} \cdot \frac{2}{7}$
$= \frac{14{,}000 \cdot 2}{1 \cdot 7}$
$= \frac{\cancel{7} \cdot 2000 \cdot 2}{1 \cdot 7} = 4000$
Carol's down payment was \$4000.

87. $\frac{3}{5} \cdot 1650 = \frac{3}{5} \cdot \frac{1650}{1}$
$= \frac{3 \cdot 1650}{5 \cdot 1} = \frac{3 \cdot \cancel{5} \cdot 330}{\cancel{5} \cdot 1}$
$= \frac{990}{1} = 990$
$1650 - 990 = 660$
John must travel 660 more miles.

89. simplified

91. $\frac{3}{8} \cdot \frac{8}{5} \cdot \frac{15}{6} = \frac{3 \cdot 8 \cdot 15}{8 \cdot 5 \cdot 6}$
$= \frac{\cancel{3} \cdot \cancel{8} \cdot 3 \cdot \cancel{5}}{\cancel{8} \cdot \cancel{5} \cdot \cancel{3} \cdot 2} = \frac{3}{2}$

93. $\frac{7}{8} \cdot \frac{4}{21} \cdot \frac{9}{2} = \frac{7 \cdot 4 \cdot 9}{8 \cdot 21 \cdot 2}$
$= \frac{\cancel{7} \cdot \cancel{2} \cdot \cancel{2} \cdot \cancel{3} \cdot 3}{\cancel{2} \cdot 4 \cdot \cancel{3} \cdot \cancel{7} \cdot \cancel{2}} = \frac{3}{4}$

95.
$$\begin{array}{r} 497 \\ 374 \\ +\ 629 \\ \hline 1500 \end{array}$$

97.
$$\begin{array}{r} 6018 \\ 376 \\ 22 \\ +\ 984 \\ \hline 7400 \end{array}$$

99.
$$\begin{array}{r} 9873 \\ +\ 3625 \\ \hline 13498 \end{array}$$

101.
$$\begin{array}{r} 3918 \\ 2457 \\ 706 \\ +\ 8415 \\ \hline 15496 \end{array}$$

103.
$$\begin{array}{r} 30{,}741 \\ 57{,}802 \\ +\ 68{,}913 \\ \hline 157{,}456 \end{array}$$

Problem Set 2.6

1. $\frac{7}{2}$

3. 4

5. $\frac{1}{7}$

7. $\frac{3}{11}$

9. $\frac{1}{4} \div \frac{5}{9} = \frac{1}{4} \cdot \frac{9}{5} = \frac{1 \cdot 9}{4 \cdot 5} = \frac{9}{20}$

11. $\frac{3}{5} \div \frac{5}{2} = \frac{3}{5} \cdot \frac{2}{5} = \frac{3 \cdot 2}{5 \cdot 5} = \frac{6}{25}$

13. $\frac{3}{8} \div \frac{5}{8} = \frac{3}{8} \cdot \frac{8}{5} = \frac{3 \cdot \cancel{8}}{\cancel{8} \cdot 5} = \frac{3}{5}$

15. $\frac{7}{5} \div \frac{3}{10} = \frac{7}{5} \cdot \frac{10}{3} = \frac{7 \cdot 10}{5 \cdot 3}$
$= \frac{7 \cdot 2 \cdot \cancel{5}}{\cancel{5} \cdot 3} = \frac{14}{3}$

17. $\frac{2}{7} \div \frac{2}{21} = \frac{2}{7} \cdot \frac{21}{2} = \frac{2 \cdot 21}{7 \cdot 2}$
$= \frac{\cancel{2} \cdot 3 \cdot \cancel{7}}{\cancel{7} \cdot \cancel{2}} = 3$

19. $\frac{5}{4} \div \frac{15}{16} = \frac{5}{4} \cdot \frac{16}{15} = \frac{5 \cdot 16}{4 \cdot 15}$
$= \frac{\cancel{5} \cdot \cancel{4} \cdot 4}{\cancel{4} \cdot 3 \cdot \cancel{5}} = \frac{4}{3}$

21. $\frac{3}{8} \div \frac{1}{8} = \frac{3}{8} \cdot \frac{8}{1} = \frac{3 \cdot \cancel{8}}{\cancel{8} \cdot 1} = 3$

Problem Set 2.6

23. $\frac{7}{10} \div \frac{1}{20} = \frac{7}{10} \bullet \frac{20}{1} = \frac{7 \bullet 20}{10 \bullet 1}$
$= \frac{7 \bullet 2 \bullet \cancel{10}}{\cancel{10} \bullet 1} = 14$

25. $\frac{5}{6} \div 5 = \frac{5}{6} \div \frac{5}{1} = \frac{5}{6} \bullet \frac{1}{5}$
$= \frac{\cancel{5} \bullet 1}{6 \bullet \cancel{5}} = \frac{1}{6}$

27. $\frac{12}{5} \div 2 = \frac{12}{5} \div \frac{2}{1} = \frac{12}{5} \bullet \frac{1}{2}$
$= \frac{12 \bullet 1}{5 \bullet 2} = \frac{\cancel{2} \bullet 6 \bullet 1}{5 \bullet \cancel{2}} = \frac{6}{5}$

29. $10 \div \frac{2}{5} = \frac{10}{1} \div \frac{2}{5} = \frac{10}{1} \bullet \frac{5}{2}$
$= \frac{10 \bullet 5}{1 \bullet 2} = \frac{\cancel{2} \bullet 5 \bullet 5}{1 \bullet \cancel{2}} = 25$

31. $40 \div \frac{8}{3} = \frac{40}{1} \div \frac{8}{3} = \frac{40}{1} \bullet \frac{3}{8}$
$= \frac{40 \bullet 3}{1 \bullet 8} = \frac{5 \bullet \cancel{8} \bullet 3}{1 \bullet \cancel{8}} = 15$

33. $\frac{6}{7} \div \frac{6}{7} = \frac{6}{7} \bullet \frac{7}{6} = \frac{\cancel{6} \bullet \cancel{7} \bullet 1}{\cancel{7} \bullet \cancel{6}} = 1$

35. $\frac{8}{5} \div \frac{5}{8} = \frac{8}{5} \bullet \frac{8}{5} = \frac{8 \bullet 8}{5 \bullet 5} = \frac{64}{25}$

37. $\frac{8}{25} \div \frac{4}{5} = \frac{8}{25} \bullet \frac{5}{4} = \frac{8 \bullet 5}{25 \bullet 4}$
$= \frac{2 \bullet \cancel{4} \bullet \cancel{5}}{5 \bullet \cancel{5} \bullet \cancel{4}} = \frac{2}{5}$

39. $\frac{15}{16} \div \frac{1}{32} = \frac{15}{16} \bullet \frac{32}{1} = \frac{15 \bullet 32}{16 \bullet 1}$
$= \frac{15 \bullet 2 \bullet \cancel{16}}{\cancel{16} \bullet 1} = \frac{30}{1} = 30$

41. $36 \div \frac{3}{4} = \frac{36}{1} \div \frac{3}{4} = \frac{36}{1} \bullet \frac{4}{3}$
$= \frac{36 \bullet 4}{1 \bullet 3} = \frac{\cancel{3} \bullet 12 \bullet 4}{1 \bullet \cancel{3}}$
$= 48$
48 pieces of crepe paper can be cut.

43. $\frac{9}{16} \div \frac{1}{32} = \frac{9}{16} \bullet \frac{32}{1} = \frac{9 \bullet 32}{16 \bullet 1}$
$= \frac{9 \bullet 2 \bullet \cancel{16}}{\cancel{16} \bullet 1} = 18$
18 bottles of shampoo can be filled.

45. $21 \div \frac{3}{5} = \frac{21}{1} \div \frac{3}{5} = \frac{21}{1} \bullet \frac{5}{3}$
$= \frac{21 \bullet 5}{1 \bullet 3} = \frac{\cancel{3} \bullet 7 \bullet 5}{1 \bullet \cancel{3}} = 35$
35 syringes can be filled.

47. $\frac{3}{4} \div \frac{3}{8} = \frac{3}{4} \bullet \frac{8}{3} = \frac{3 \bullet 8}{4 \bullet 3}$
$= \frac{\cancel{3} \bullet 2 \bullet \cancel{4}}{\cancel{4} \bullet \cancel{3}} = 2$
Two pieces of fabric can be cut.

49. $2040 \div \frac{3}{4} = \frac{2040}{1} \div \frac{3}{4}$
$= \frac{2040}{1} \bullet \frac{4}{3} = \frac{2040 \bullet 4}{1 \bullet 3}$
$= \frac{\cancel{3} \bullet 680 \bullet 4}{1 \bullet \cancel{3}} = 2720$
The capacity of the stadium is 2720 people.

51. $3 \bullet \frac{7}{8} = \frac{3}{1} \bullet \frac{7}{8} = \frac{3 \bullet 7}{1 \bullet 8} = \frac{21}{8}$
$\frac{21}{8} \div \frac{1}{16} = \frac{21}{8} \bullet \frac{16}{1} = \frac{21 \bullet 16}{8 \bullet 1}$
$= \frac{21 \bullet 2 \bullet \cancel{8}}{\cancel{8} \bullet 1} = 42$
Wesley can make 42 servings of nuts.

53. reciprocal

55. dividend, divisor

57. $\frac{165}{74} \div \frac{25}{88} = \frac{165}{74} \bullet \frac{88}{25}$
$= \frac{165 \bullet 88}{74 \bullet 25}$
$= \frac{3 \bullet \cancel{5} \bullet 11 \bullet \cancel{2} \bullet 44}{\cancel{2} \bullet 37 \bullet \cancel{5} \bullet 5}$
$= \frac{1452}{185}$

59. $\begin{array}{r} 932 \\ -\underline{493} \\ 439 \end{array}$

61. $\begin{array}{r} 804 \\ -\underline{632} \\ 172 \end{array}$

63. $\begin{array}{r} 6382 \\ -\underline{3491} \\ 2891 \end{array}$

65. $\begin{array}{r} 6003 \\ -\underline{3974} \\ 2029 \end{array}$

67. $\begin{array}{r} 95{,}000 \\ -\underline{48{,}016} \\ 46{,}984 \end{array}$

Chapter 2 Additional Exercises

1. proper fractions $= \frac{3}{5}, \frac{7}{9}$
   improper fractions
   $= \frac{5}{5}, \frac{9}{8}, \frac{11}{4}$

3. proper fractions $= \frac{5}{12}, \frac{3}{4}$
   improper fractions
   $= \frac{15}{7}, \frac{8}{5}, \frac{7}{7}$

5. $\frac{38}{38} = 1$

7. $\frac{55}{1} = 55$

9. 6 is composite because
   $6 = 3 \bullet 2$

11. 31 is prime

13. $25 = 5 \bullet 5$

15. $32 = 2 \bullet 2 \bullet 2 \bullet 2 \bullet 2$

17. $28 = 2 \bullet 2 \bullet 7$

19. $78 = 2 \bullet 3 \bullet 13$

21. 42, 616, 15128, 3804, 4800

23. 735, 4800, 835

25. $\frac{1}{7} \bullet \frac{2}{3} = \frac{1 \bullet 2}{7 \bullet 3} = \frac{2}{21}$

27. $\frac{8}{7} \bullet \frac{3}{5} = \frac{8 \bullet 3}{7 \bullet 5} = \frac{24}{35}$

29. $\frac{1}{40} \bullet \frac{1}{39} = \frac{1 \bullet 1}{40 \bullet 39} = \frac{1}{1560}$

31. $\frac{11}{40} \bullet \frac{77}{3} = \frac{11 \bullet 77}{40 \bullet 3} = \frac{847}{120}$

33. No, since $5 \bullet 36 \neq 6 \bullet 20$
   $180 \neq 120$

35. No, since $8 \bullet 20 \neq 5 \bullet 33$
   $160 \neq 165$

37. $18 \bullet \frac{1}{3} = \frac{18}{1} \bullet \frac{1}{3} = \frac{18 \bullet 1}{1 \bullet 3}$
   $= \frac{\cancel{3} \bullet 6 \bullet 1}{1 \bullet \cancel{3}} = 6$

39. $\frac{3}{7} \bullet \frac{14}{9} = \frac{3 \bullet 14}{7 \bullet 9}$
   $= \frac{\cancel{3} \bullet 2 \bullet \cancel{7}}{\cancel{7} \bullet \cancel{3} \bullet 3} = \frac{2}{3}$

41. $\frac{4}{5} \bullet \frac{5}{4} = \frac{\cancel{4} \bullet \cancel{5} \bullet 1}{\cancel{5} \bullet \cancel{4} \bullet 1} = 1$

43. $\frac{2}{3} \bullet 27 = \frac{2}{3} \bullet \frac{27}{1} = \frac{2 \bullet 27}{3 \bullet 1}$
   $= \frac{2 \bullet \cancel{3} \bullet 9}{\cancel{3} \bullet 1} = 18$

45. $\frac{7}{10} \bullet \frac{40}{21} = \frac{7 \bullet 40}{10 \bullet 21}$
   $= \frac{\cancel{7} \bullet 4 \bullet \cancel{10}}{\cancel{10} \bullet 3 \bullet \cancel{7}} = \frac{4}{3}$

47. $16 \bullet \frac{1}{16} = \frac{16}{1} \bullet \frac{1}{16}$
   $= \frac{\cancel{16} \bullet 1}{1 \bullet \cancel{16}} = 1$

49. $\frac{7}{8} \bullet \frac{3}{8} = \frac{7 \bullet 3}{8 \bullet 8} = \frac{21}{64}$
   Linda used $\frac{21}{64}$ of the paper.

51. $\frac{3}{7} \bullet \frac{5}{6} \bullet \frac{14}{25} = \frac{3 \bullet 5 \bullet 14}{7 \bullet 6 \bullet 25}$
   $= \frac{\cancel{3} \bullet \cancel{5} \bullet \cancel{2} \bullet \cancel{7} \bullet 1}{\cancel{7} \bullet \cancel{2} \bullet \cancel{3} \bullet \cancel{5} \bullet 5} = \frac{1}{5}$

53. $\frac{12}{11}$

55. $\frac{1}{6}$

57. $\frac{7}{8} \div \frac{3}{2} = \frac{7}{8} \bullet \frac{2}{3} = \frac{7 \bullet 2}{8 \bullet 3}$
   $= \frac{7 \bullet \cancel{2}}{\cancel{2} \bullet 4 \bullet 3} = \frac{7}{12}$

Chapter 2 Additional Exercises

59. $\frac{6}{15} \div \frac{3}{5} = \frac{6}{15} \bullet \frac{5}{3} = \frac{6 \bullet 5}{15 \bullet 3}$
$= \frac{2 \bullet \cancel{3} \bullet \cancel{5}}{\cancel{3} \bullet \cancel{5} \bullet 3} = \frac{2}{3}$

61. $\frac{7}{25} \div \frac{14}{15} = \frac{7}{25} \bullet \frac{15}{14} = \frac{7 \bullet 15}{25 \bullet 14}$
$= \frac{\cancel{7} \bullet 3 \bullet \cancel{5}}{\cancel{5} \bullet 5 \bullet 2 \bullet \cancel{7}} = \frac{3}{10}$

63. $\frac{44}{51} \div \frac{33}{102} = \frac{44}{51} \bullet \frac{102}{33}$
$= \frac{44 \bullet 102}{51 \bullet 33}$
$= \frac{4 \bullet \cancel{11} \bullet 2 \bullet \cancel{51}}{\cancel{51} \bullet 3 \bullet \cancel{11}} = \frac{8}{3}$

65. $6 \div \frac{3}{8} = \frac{6}{1} \div \frac{3}{8} = \frac{6}{1} \bullet \frac{8}{3}$
$= \frac{6 \bullet 8}{1 \bullet 3} = \frac{2 \bullet \cancel{3} \bullet 8}{1 \bullet \cancel{3}} = 16$
One pound of cake costs \$16.

Chapter 2 Practice Test

1. $\frac{27}{27} = 1$

3. $\frac{8}{1} = 8$

5.
```
  227
3)682
  6
  08
   6
   22
   21
    1
```
682 is not divisible by 3 because there is a remainder.

7. $\frac{3}{7} \bullet \frac{8}{5} = \frac{3 \bullet 8}{7 \bullet 5} = \frac{24}{35}$

9. $\frac{3}{15} = \frac{1 \bullet \cancel{3}}{\cancel{3} \bullet 5} = \frac{1}{5}$

11. $\frac{110}{760} = \frac{\cancel{10} \bullet 11}{\cancel{10} \bullet 76} = \frac{11}{76}$

13. No, since $9 \bullet 16 \neq 5 \bullet 27$
$144 \neq 135$

15. $15 \bullet \frac{2}{5} = \frac{15}{1} \bullet \frac{2}{5} = \frac{15 \bullet 2}{1 \bullet 5}$
$= \frac{3 \bullet \cancel{5} \bullet 2}{1 \bullet \cancel{5}} = 6$

17. $\frac{8}{7} \bullet \frac{21}{16} = \frac{8 \bullet 21}{7 \bullet 16}$
$= \frac{\cancel{8} \bullet 3 \bullet \cancel{7}}{\cancel{7} \bullet 2 \bullet \cancel{8}} = \frac{3}{2}$

19. 12

21. $\frac{5}{8} \div \frac{25}{18} = \frac{5}{8} \bullet \frac{18}{25} = \frac{5 \bullet 18}{8 \bullet 25}$
$= \frac{\cancel{5} \bullet \cancel{2} \bullet 9}{\cancel{2} \bullet 4 \bullet \cancel{5} \bullet 5} = \frac{9}{20}$

23. $18 \div \frac{9}{2} = \frac{18}{1} \div \frac{9}{2} = \frac{18}{1} \bullet \frac{2}{9}$
$= \frac{18 \bullet 2}{1 \bullet 9} = \frac{2 \bullet \cancel{9} \bullet 2}{1 \bullet \cancel{9}} = 4$

25. $16 \bullet \frac{3}{4} = \frac{16}{1} \bullet \frac{3}{4} = \frac{16 \bullet 3}{1 \bullet 4}$
$= \frac{\cancel{4} \bullet 4 \bullet 3}{1 \bullet \cancel{4}} = 12$
Anton needs 12 teaspoons of salt.

Chapters 1 and 2 Cumulative Review

1. ten thousands

3.
```
  713
+ 895
 1608
```

5.
```
 8614
-3982
 4632
```

7.
```
  85 ≈   90
+324 ≈ +320
        410
```

9.
```
  76
× 34
 304
228
2584
```

Chapters 1 and 2 Cumulative Review

11.
$$\begin{array}{r} 71 \\ 8\overline{)575} \\ \underline{56} \\ 15 \\ \underline{8} \\ 7 \end{array}$$
The answer is 71 R 7.

13.
$$\begin{array}{r} 36,003 \\ \underline{-27,345} \\ 8,658 \end{array}$$
State University has 8,658 more students than City Community College.

15.
$$\begin{array}{r} 274 \\ \underline{\times \quad 20} \\ 5480 \end{array}$$
There are 5480 nickels in $274.

17. $72 = 2 \bullet 2 \bullet 2 \bullet 3 \bullet 3$

19. $\frac{0}{3} = 0$

21. $\frac{24}{32} = \frac{3 \bullet \cancel{8}}{4 \bullet \cancel{8}} = \frac{3}{4}$

23. $\frac{3}{5} \bullet 30 = \frac{3}{5} \bullet \frac{30}{1} = \frac{3 \bullet 30}{5 \bullet 1}$
$= \frac{3 \bullet \cancel{5} \bullet 6}{\cancel{5} \bullet 1} = 18$

25. $\frac{4}{9}$

27. $\frac{5}{6} \div \frac{35}{12} = \frac{5}{6} \bullet \frac{12}{35} = \frac{5 \bullet 12}{6 \bullet 35}$
$= \frac{\cancel{5} \bullet 2 \bullet \cancel{6}}{\cancel{6} \bullet \cancel{5} \bullet 7} = \frac{2}{7}$

29. $\frac{5}{1} \div \frac{10}{3} = \frac{5}{1} \bullet \frac{3}{10} = \frac{5 \bullet 3}{1 \bullet 10}$
$= \frac{\cancel{5} \bullet 3}{1 \bullet 2 \bullet \cancel{5}} = \frac{3}{2}$

31. $\frac{16500}{1} \bullet \frac{1}{5} = \frac{16500 \bullet 1}{1 \bullet 5}$
$= \frac{\cancel{5} \bullet 3300 \bullet 1}{1 \bullet \cancel{5}} = 3300$
Karl's down payment was $3300.

Problem Set 3.1

1. $\frac{2}{6} + \frac{3}{6} = \frac{2+3}{6} = \frac{5}{6}$

3. $\frac{1}{9} + \frac{2}{9} = \frac{1+2}{9} = \frac{3}{9} = \frac{1}{3}$

5. $\frac{7}{8} + \frac{5}{8} = \frac{7+5}{8} = \frac{12}{8} = \frac{3}{2}$

7. $\frac{7}{6} + \frac{2}{6} = \frac{7+2}{6} = \frac{9}{6} = \frac{3}{2}$

9. $\frac{3}{2} + \frac{7}{2} = \frac{3+7}{2} = \frac{10}{2} = 5$

11. $\frac{3}{10} + \frac{4}{10} = \frac{3+4}{10} = \frac{7}{10}$

13. $\frac{3}{8} + \frac{5}{8} = \frac{3+5}{8} = \frac{8}{8} = 1$
Judson bought 1 pound of coffee.

15. $\frac{5}{9} - \frac{1}{9} = \frac{5-1}{9} = \frac{4}{9}$

17. $\frac{9}{10} - \frac{4}{10} = \frac{9-4}{10} = \frac{5}{10} = \frac{1}{2}$

19. $\frac{7}{5} - \frac{2}{5} = \frac{7-2}{5} = \frac{5}{5} = 1$

21. $\frac{17}{12} - \frac{9}{12} = \frac{17-9}{12} = \frac{8}{12} = \frac{2}{3}$

23. $\frac{7}{3} - \frac{5}{3} = \frac{7-5}{3} = \frac{2}{3}$

25. $\frac{17}{6} - \frac{8}{6} = \frac{17-8}{6} = \frac{9}{6} = \frac{3}{2}$

27. $\frac{9}{10} - \frac{6}{10} = \frac{9-6}{10} = \frac{3}{10}$
Shelly bought $\frac{3}{10}$ pound more hamburger than pork.

29. like

31. $\frac{2}{7} + \frac{3}{7} + \frac{5}{7} = \frac{2+3+5}{7} = \frac{10}{7}$

33. $\frac{29}{45} + \frac{16}{45} + \frac{3}{45} = \frac{29+16+3}{45}$
$= \frac{48}{45} = \frac{16}{15}$

35. $18 = 2 \bullet 3 \bullet 3$

37. $12 = 2 \bullet 2 \bullet 3$

Problem Set 3.1

39. $20 = 2 \bullet 2 \bullet 5$

41. $50 = 2 \bullet 5 \bullet 5$

Problem Set 3.2

1. $4 = 2 \bullet 2$
$8 = 2 \bullet 2 \bullet 2$
$\text{LCD} = 2 \bullet 2 \bullet 2 = 8$

3. $5 = 5$
$15 = 3 \bullet 5$
$\text{LCD} = 3 \bullet 5 = 15$

5. $3 = 3$
$4 = 2 \bullet 2$
$\text{LCD} = 2 \bullet 2 \bullet 3 = 12$

7. $12 = 2 \bullet 2 \bullet 3$
$16 = 2 \bullet 2 \bullet 2 \bullet 2$
$\text{LCD} = 2 \bullet 2 \bullet 2 \bullet 2 \bullet 3 = 48$

9. $9 = 3 \bullet 3$
$12 = 2 \bullet 2 \bullet 3$
$\text{LCD} = 2 \bullet 2 \bullet 3 \bullet 3 = 36$

11. $40 = 2 \bullet 2 \bullet 2 \bullet 5$
$50 = 2 \bullet 5 \bullet 5$
$\text{LCD} = 2 \bullet 2 \bullet 2 \bullet 5 \bullet 5 = 200$

13. $35 = 5 \bullet 7$
$55 = 5 \bullet 11$
$\text{LCD} = 5 \bullet 7 \bullet 11 = 385$

15. $22 = 2 \bullet 11$
$33 = 3 \bullet 11$
$\text{LCD} = 2 \bullet 3 \bullet 11 = 66$

17. $2 = 2$
$8 = 2 \bullet 2 \bullet 2$
$12 = 2 \bullet 2 \bullet 3$
$\text{LCD} = 2 \bullet 2 \bullet 2 \bullet 3 = 24$

19. $14 = 2 \bullet 7$
$21 = 3 \bullet 7$
$28 = 2 \bullet 2 \bullet 7$
$\text{LCD} = 2 \bullet 2 \bullet 3 \bullet 7 = 84$

21. $9 = 3 \bullet 3$
$36 = 2 \bullet 2 \bullet 3 \bullet 3$
$\text{LCD} = 2 \bullet 2 \bullet 3 \bullet 3 = 36$

23. $11 = 11$
$44 = 2 \bullet 2 \bullet 11$
$\text{LCD} = 2 \bullet 2 \bullet 11 = 44$

25. $6 = 2 \bullet 3$
$7 = 7$
$\text{LCD} = 2 \bullet 3 \bullet 7 = 42$

27. $5 = 5$
$14 = 2 \bullet 7$
$\text{LCD} = 2 \bullet 5 \bullet 7 = 70$

29. $3 = 3$
$5 = 5$
$7 = 7$
$\text{LCD} = 3 \bullet 5 \bullet 7 = 105$

31. $4 = 2 \bullet 2$
$7 = 7$
$11 = 11$
$\text{LCD} = 2 \bullet 2 \bullet 7 \bullet 11 = 308$

33. Least common denominator

35. $16 = 2 \bullet 2 \bullet 2 \bullet 2$
$64 = 2 \bullet 2 \bullet 2 \bullet 2 \bullet 2 \bullet 2$
$12 = 2 \bullet 2 \bullet 3$
$\text{LCD} = 2 \bullet 2 \bullet 2 \bullet 2 \bullet 2 \bullet 2 \bullet 3 = 192$

37. $27 = 3 \bullet 3 \bullet 3$
$36 = 2 \bullet 2 \bullet 3 \bullet 3$
$54 = 2 \bullet 3 \bullet 3 \bullet 3$
$\text{LCD} = 2 \bullet 2 \bullet 3 \bullet 3 \bullet 3 = 108$

39. $\frac{4}{6} = \frac{\cancel{2} \bullet 2}{\cancel{2} \bullet 3} = \frac{2}{3}$

41. $\frac{15}{25} = \frac{3 \bullet \cancel{5}}{5 \bullet \cancel{5}} = \frac{3}{5}$

43. $\frac{3}{8} \bullet \frac{16}{5} = \frac{3 \bullet 16}{8 \bullet 5}$
$= \frac{3 \bullet 2 \bullet \cancel{8}}{\cancel{8} \bullet 5} = \frac{6}{5}$

45. $\frac{9}{7} \bullet \frac{14}{15} = \frac{9 \bullet 14}{7 \bullet 15}$
$= \frac{\cancel{3} \bullet 3 \bullet 2 \bullet \cancel{7}}{\cancel{7} \bullet \cancel{3} \bullet 5} = \frac{6}{5}$

47. $\frac{8}{33} \bullet \frac{55}{14} = \frac{8 \bullet 55}{33 \bullet 14}$
$= \frac{\cancel{2} \bullet 4 \bullet 5 \bullet \cancel{11}}{3 \bullet \cancel{11} \bullet \cancel{2} \bullet 7} = \frac{20}{21}$

Problem Set 3.3

1. $\frac{1}{2} \bullet \frac{2}{2} = \frac{2}{4}$

3. $\frac{2}{3} \bullet \frac{2}{2} = \frac{4}{6}$

5. $\frac{3}{5} \bullet \frac{5}{5} = \frac{15}{25}$

Problem Set 3.3

7. $\frac{6}{7} \bullet \frac{6}{6} = \frac{36}{42}$

9. $\frac{9}{4} \bullet \frac{4}{4} = \frac{36}{16}$

11. $\frac{8}{15} \bullet \frac{7}{7} = \frac{56}{105}$

13. $\frac{1}{4} + \frac{3}{8} = \frac{1}{4} \bullet \frac{2}{2} + \frac{3}{8}$
$= \frac{2}{8} + \frac{3}{8} = \frac{5}{8}$

15. $\frac{3}{7} + \frac{1}{3} = \frac{3}{7} \bullet \frac{3}{3} + \frac{1}{3} \bullet \frac{7}{7}$
$= \frac{9}{21} + \frac{7}{21} = \frac{16}{21}$

17. $\frac{3}{8} + \frac{5}{6} = \frac{3}{8} \bullet \frac{3}{3} + \frac{5}{6} \bullet \frac{4}{4}$
$= \frac{9}{24} + \frac{20}{24} = \frac{29}{24}$

19. $\frac{3}{4} + \frac{9}{16} = \frac{3}{4} \bullet \frac{4}{4} + \frac{9}{16}$
$= \frac{12}{16} + \frac{9}{16} = \frac{21}{16}$

21. $\frac{1}{5} + \frac{8}{3} = \frac{1}{5} \bullet \frac{3}{3} + \frac{8}{3} \bullet \frac{5}{5}$
$= \frac{3}{15} + \frac{40}{15} = \frac{43}{15}$

23. $\frac{7}{6} + \frac{3}{4} = \frac{7}{6} \bullet \frac{2}{2} + \frac{3}{4} \bullet \frac{3}{3}$
$= \frac{14}{12} + \frac{9}{12} = \frac{23}{12}$

25. $\frac{7}{10} + \frac{1}{100} = \frac{7}{10} \bullet \frac{10}{10} + \frac{1}{100}$
$= \frac{70}{100} + \frac{1}{100} = \frac{71}{100}$

27. $\frac{7}{12} + \frac{3}{16} = \frac{7}{12} \bullet \frac{4}{4} + \frac{3}{16} \bullet \frac{3}{3}$
$= \frac{28}{48} + \frac{9}{48} = \frac{37}{48}$

29. $\frac{5}{12} + \frac{3}{8} = \frac{5}{12} \bullet \frac{2}{2} + \frac{3}{8} \bullet \frac{3}{3}$
$= \frac{10}{24} + \frac{9}{24} = \frac{19}{24}$

31. $\frac{9}{10} + \frac{7}{100} = \frac{9}{10} \bullet \frac{10}{10} + \frac{7}{100}$
$= \frac{90}{100} + \frac{7}{100} = \frac{97}{100}$

33. $\frac{5}{8} + \frac{4}{9} = \frac{5}{8} \bullet \frac{9}{9} + \frac{4}{9} \bullet \frac{8}{8}$
$= \frac{45}{72} + \frac{32}{72} = \frac{77}{72}$

35. $\frac{3}{16} + \frac{5}{24} = \frac{3}{16} \bullet \frac{3}{3} + \frac{5}{24} \bullet \frac{2}{2}$
$= \frac{9}{48} + \frac{10}{48} = \frac{19}{48}$

37. $\frac{1}{3} + \frac{4}{5} + \frac{1}{6}$
$= \frac{1}{3} \bullet \frac{10}{10} + \frac{4}{5} \bullet \frac{6}{6} + \frac{1}{6} \bullet \frac{5}{5}$
$= \frac{10}{30} + \frac{24}{30} + \frac{5}{30}$
$= \frac{39}{30} = \frac{13}{10}$

39. $\frac{1}{10} + \frac{3}{100} + \frac{11}{1000}$
$= \frac{1}{10} \bullet \frac{100}{100} + \frac{3}{100}$
$\bullet \frac{10}{10} + \frac{11}{1000}$
$= \frac{100}{1000} + \frac{30}{1000} + \frac{11}{1000}$
$= \frac{141}{1000}$

41. $\frac{5}{32} + \frac{3}{16} + \frac{7}{48}$
$= \frac{5}{32} \bullet \frac{3}{3} + \frac{3}{16} \bullet \frac{6}{6}$
$+ \frac{7}{48} \bullet \frac{2}{2}$
$= \frac{15}{96} + \frac{18}{96} + \frac{14}{96} = \frac{47}{96}$

43. $\frac{2}{3} - \frac{1}{7} = \frac{2}{3} \bullet \frac{7}{7} - \frac{1}{7} \bullet \frac{3}{3}$
$= \frac{14}{21} - \frac{3}{21} = \frac{11}{21}$

45. $\frac{9}{8} - \frac{3}{16} = \frac{9}{8} \bullet \frac{2}{2} - \frac{3}{16}$
$= \frac{18}{16} - \frac{3}{16} = \frac{15}{16}$

Problem Set 3.3

47. $\frac{7}{8}-\frac{5}{6} = \frac{7}{8}\bullet\frac{3}{3}-\frac{5}{6}\bullet\frac{4}{4}$
$= \frac{21}{24}-\frac{20}{24} = \frac{1}{24}$

49. $\frac{8}{7}-\frac{3}{14} = \frac{8}{7}\bullet\frac{2}{2}-\frac{3}{14}$
$= \frac{16}{14}-\frac{3}{14} = \frac{13}{14}$

51. $\frac{7}{18}-\frac{1}{12} = \frac{7}{18}\bullet\frac{2}{2}-\frac{1}{12}\bullet\frac{3}{3}$
$= \frac{14}{36}-\frac{3}{36} = \frac{11}{36}$

53. $\frac{7}{10}-\frac{3}{100} = \frac{7}{10}\bullet\frac{10}{10}-\frac{3}{100}$
$= \frac{70}{100}-\frac{3}{100} = \frac{67}{100}$

55. $\frac{11}{45}-\frac{1}{25} = \frac{11}{45}\bullet\frac{5}{5}-\frac{1}{25}\bullet\frac{9}{9}$
$= \frac{55}{225}-\frac{9}{225} = \frac{46}{225}$

57. $\frac{14}{15}-\frac{3}{10} = \frac{14}{15}\bullet\frac{2}{2}-\frac{3}{10}\bullet\frac{3}{3}$
$= \frac{28}{30}-\frac{9}{30} = \frac{19}{30}$

59. $\frac{1}{2}+\frac{2}{3} = \frac{1}{2}\bullet\frac{3}{3}+\frac{2}{3}\bullet\frac{2}{2}$
$= \frac{3}{6}+\frac{4}{6} = \frac{7}{6}$
Ken bought $\frac{7}{6}$ pounds of coffee.

61. $\frac{7}{8}+\frac{3}{4} = \frac{7}{8}+\frac{3}{4}\bullet\frac{2}{2}$
$= \frac{7}{8}+\frac{6}{8} = \frac{13}{8}$
Mai had $\frac{13}{8}$ yards of lace.

63. $\frac{3}{5}+\frac{7}{10} = \frac{3}{5}\bullet\frac{2}{2}+\frac{7}{10}$
$= \frac{6}{10}+\frac{7}{10} = \frac{13}{10}$
The chemist had $\frac{13}{10}$ liters of solution.

65. $\frac{9}{10}-\frac{3}{5} = \frac{9}{10}-\frac{3}{5}\bullet\frac{2}{2}$
$= \frac{9}{10}-\frac{6}{10} = \frac{3}{10}$
Jim walked $\frac{3}{10}$ mile farther than Joan.

67. $\frac{3}{4}-\frac{1}{3} = \frac{3}{4}\bullet\frac{3}{3}-\frac{1}{3}\bullet\frac{4}{4}$
$= \frac{9}{12}-\frac{4}{12} = \frac{5}{12}$
The stew requires $\frac{5}{12}$ pound more beef than the casserole.

69. unlike

71. $\frac{1}{4}+\frac{1}{8}+\frac{1}{16}$
$= \frac{1}{4}\bullet\frac{4}{4}+\frac{1}{8}\bullet\frac{2}{2}+\frac{1}{16}$
$= \frac{4}{16}+\frac{2}{16}+\frac{1}{16} = \frac{7}{16}$
$1-\frac{7}{16} = \frac{16}{16}-\frac{7}{16} = \frac{9}{16}$
The fourth roommate is left with $\frac{9}{16}$ of the pizza.

73. proper fractions $= \frac{3}{4}, \frac{1}{3}$
improper fractions
$= \frac{8}{5}, \frac{7}{5}, \frac{8}{8}$

75. proper fractions $= \frac{5}{6}, \frac{1}{5}, \frac{11}{12}$
improper fractions $= \frac{10}{9}, \frac{4}{4}$

77. $\frac{0}{28} = 0$

79. $\frac{66}{66} = 1$

81. $\frac{0}{9} = 0$

Problem Set 3.4

1. $5\frac{1}{2} = \frac{(2 \times 5)+1}{2} = \frac{11}{2}$

Problem Set 3.4

3. $2\frac{1}{4} = \frac{(4 \times 2) + 1}{4} = \frac{9}{4}$

5. $6\frac{1}{8} = \frac{(8 \times 6) + 1}{8} = \frac{49}{8}$

7. $9\frac{3}{7} = \frac{(7 \times 9) + 3}{7} = \frac{66}{7}$

9. $10\frac{3}{4} = \frac{(4 \times 10) + 3}{4} = \frac{43}{4}$

11. $15\frac{1}{4} = \frac{(4 \times 15) + 1}{4} = \frac{61}{4}$

13. $4\frac{5}{8} = \frac{(8 \times 4) + 5}{8} = \frac{37}{8}$

15. $1\frac{5}{9} = \frac{(9 \times 1) + 5}{9} = \frac{14}{9}$

17. $5\frac{3}{10} = \frac{(10 \times 5) + 3}{10} = \frac{53}{10}$

19. $12\frac{5}{6} = \frac{(6 \times 12) + 5}{6} = \frac{77}{6}$

21. $3\frac{1}{12} = \frac{(12 \times 3) + 1}{12} = \frac{37}{12}$

23. $6\frac{3}{100} = \frac{(100 \times 6) + 3}{100} = \frac{603}{100}$

25. $40\frac{3}{4} = \frac{(4 \times 40) + 3}{4} = \frac{163}{4}$

27. $3\frac{27}{40} = \frac{(40 \times 3) + 27}{40} = \frac{147}{40}$

29. $$\begin{array}{r} 1 \\ 2\overline{)3} \\ \underline{2} \\ 1 \end{array} \qquad \frac{3}{2} = 1\frac{1}{2}$$

31. $$\begin{array}{r} 1 \\ 5\overline{)9} \\ \underline{5} \\ 4 \end{array} \qquad \frac{9}{5} = 1\frac{4}{5}$$

33. $$\begin{array}{r} 3 \\ 4\overline{)15} \\ \underline{12} \\ 3 \end{array} \qquad \frac{15}{4} = 3\frac{3}{4}$$

35. $$\begin{array}{r} 2 \\ 8\overline{)21} \\ \underline{16} \\ 5 \end{array} \qquad \frac{21}{8} = 2\frac{5}{8}$$

37. $$\begin{array}{r} 9 \\ 7\overline{)64} \\ \underline{63} \\ 1 \end{array} \qquad \frac{64}{7} = 9\frac{1}{7}$$

39. $$\begin{array}{r} 7 \\ 4\overline{)30} \\ \underline{28} \\ 2 \end{array} \qquad \frac{30}{4} = 7\frac{2}{4} = 7\frac{1}{2}$$

41. $$\begin{array}{r} 10 \\ 5\overline{)51} \\ \underline{5\ \,} \\ 01 \end{array} \qquad \frac{51}{5} = 10\frac{1}{5}$$

43. $$\begin{array}{r} 1 \\ 9\overline{)15} \\ \underline{9} \\ 6 \end{array} \qquad \frac{15}{9} = 1\frac{6}{9} = 1\frac{2}{3}$$

45. $$\begin{array}{r} 8 \\ 100\overline{)823} \\ \underline{800} \\ 23 \end{array} \qquad \frac{823}{100} = 8\frac{23}{100}$$

47. $$\begin{array}{r} 32 \\ 7\overline{)228} \\ \underline{21\ \,} \\ 18 \\ \underline{14} \\ 4 \end{array} \qquad \frac{228}{7} = 32\frac{4}{7}$$

49. $$\begin{array}{r} 5 \\ 5\overline{)28} \\ \underline{25} \\ 3 \end{array} \qquad 5\frac{3}{5}$$

51. $$\begin{array}{r} 191 \\ 3\overline{)574} \\ \underline{3\ \ \ \,} \\ 27\ \, \\ \underline{27\ \,} \\ 04 \\ \underline{3} \\ 1 \end{array} \qquad 191\frac{1}{3}$$

53. $$\begin{array}{r} 2610 \\ 3\overline{)7831} \\ \underline{6\ \ \ \ \ \,} \\ 18\ \ \ \, \\ \underline{18\ \ \ \,} \\ 03\ \, \\ \underline{3\ \,} \\ 01 \end{array} \qquad 2610\frac{1}{3}$$

Problem Set 3.4

55. $$\begin{array}{r} 42 \\ 23\overline{)984} \\ \underline{92} \\ 64 \\ \underline{46} \\ 18 \end{array} \qquad 42\frac{18}{23}$$

57. mixed

59. mixed

61. $$\begin{array}{r} 91 \\ 98\overline{)9014} \\ \underline{882} \\ 194 \\ \underline{98} \\ 96 \end{array} \qquad 91\frac{96}{98} = 91\frac{48}{49}$$

63. $$\begin{array}{r} 69 \\ 104\overline{)7185} \\ \underline{624} \\ 945 \\ \underline{936} \\ 9 \end{array} \qquad 69\frac{9}{104}$$

65. $\frac{5}{6} \bullet \frac{5}{8} = \frac{5 \bullet 5}{6 \bullet 8} = \frac{25}{48}$

67. $\frac{5}{4} \bullet \frac{15}{7} = \frac{5 \bullet 15}{4 \bullet 7} = \frac{75}{28}$

69. $\frac{12}{7} \bullet \frac{22}{5} = \frac{12 \bullet 22}{7 \bullet 5} = \frac{264}{35}$

71. $\frac{9}{5} \bullet \frac{46}{1} = \frac{9 \bullet 46}{5 \bullet 1} = \frac{414}{5}$

73. $\frac{57}{1} \bullet \frac{3}{8} = \frac{57 \bullet 3}{1 \bullet 8} = \frac{171}{8}$

Problem Set 3.5

1. $4\frac{1}{3} \bullet \frac{3}{5} = \frac{13}{\cancel{3}} \bullet \frac{\cancel{3}}{5} = \frac{13}{5} = 2\frac{3}{5}$

3. $6 \bullet 2\frac{1}{4} = \frac{6}{1} \bullet \frac{9}{4} = \frac{\cancel{2} \bullet 3}{1} \bullet \frac{9}{\cancel{2} \bullet 2}$
$= \frac{27}{2} = 13\frac{1}{2}$

5. $3\frac{1}{2} \bullet 4\frac{1}{3} = \frac{7}{2} \bullet \frac{13}{3} = \frac{91}{6} = 15\frac{1}{6}$

7. $5\frac{1}{5} \bullet 2\frac{1}{2} = \frac{26}{\cancel{5}} \bullet \frac{\cancel{5}}{2} = \frac{26}{2} = 13$

9. $5\frac{1}{7} \bullet 4\frac{1}{4} = \frac{36}{7} \bullet \frac{17}{4}$
$= \frac{9 \bullet \cancel{4}}{7} \bullet \frac{17}{\cancel{4}}$
$= \frac{153}{7} = 21\frac{6}{7}$

11. $12\frac{1}{2} \bullet 3\frac{1}{4} = \frac{25}{2} \bullet \frac{13}{4}$
$= \frac{325}{8} = 40\frac{5}{8}$

13. $3\frac{1}{3} \div \frac{3}{2} = \frac{10}{3} \div \frac{3}{2} = \frac{10}{3} \bullet \frac{2}{3}$
$= \frac{20}{9} = 2\frac{2}{9}$

15. $6 \div 3\frac{1}{3} = \frac{6}{1} \div \frac{10}{3} = \frac{6}{1} \bullet \frac{3}{10}$
$= \frac{\cancel{2} \bullet 3}{1} \bullet \frac{3}{\cancel{2} \bullet 5}$
$= \frac{9}{5} = 1\frac{4}{5}$

17. $\frac{3}{8} \div 2\frac{1}{2} = \frac{3}{8} \div \frac{5}{2} = \frac{3}{8} \bullet \frac{2}{5}$
$= \frac{3}{\cancel{2} \bullet 4} \bullet \frac{\cancel{2}}{5} = \frac{3}{20}$

19. $5\frac{7}{8} \div 2 = \frac{47}{8} \div \frac{2}{1} = \frac{47}{8} \bullet \frac{1}{2}$
$= \frac{47}{16} = 2\frac{15}{16}$

21. $4\frac{1}{5} \div 1\frac{2}{3} = \frac{21}{5} \div \frac{5}{3} = \frac{21}{5} \bullet \frac{3}{5}$
$= \frac{63}{25} = 2\frac{13}{25}$

23. $6\frac{1}{2} \div 5\frac{1}{4} = \frac{13}{2} \div \frac{21}{4}$
$= \frac{13}{2} \bullet \frac{4}{21} = \frac{13}{\cancel{2}} \bullet \frac{\cancel{2} \bullet 2}{21}$
$= \frac{26}{21} = 1\frac{5}{21}$

25. $21 \bullet 5\frac{2}{3} = \frac{21}{1} \bullet \frac{17}{3}$
$= \frac{\cancel{3} \bullet 7}{1} \bullet \frac{17}{\cancel{3}} = 119$

Scott's Chevrolet travels 119 miles on $5\frac{2}{3}$ gallons of gasoline.

Problem Set 3.5

27. $7 \bullet 8\frac{1}{4} = \frac{7}{1} \bullet \frac{33}{4} = \frac{231}{4} = 57\frac{3}{4}$

Manuel earns $\$57\frac{3}{4}$ for working $8\frac{1}{4}$ hours.

29. $4\frac{1}{2} \bullet 125 = \frac{9}{2} \bullet \frac{125}{1}$

$= \frac{1125}{2} = 562\frac{1}{2}$

Susan spends \$562.50 on her house payment.

31. $410 \bullet 3\frac{5}{8} = \frac{410}{1} \bullet \frac{29}{8}$

$= \frac{\cancel{2} \bullet 205}{1} \bullet \frac{29}{\cancel{2} \bullet 4}$

$= \frac{5945}{4} = 1486\frac{1}{4}$

It will cost \$1,486.25.

33. $36,000 \div 2\frac{1}{4} = \frac{36000}{1} \div \frac{9}{4}$

$= \frac{36000}{1} \bullet \frac{4}{9}$

$= \frac{\cancel{9} \bullet 4000}{1} \bullet \frac{4}{\cancel{9}} = 16000$

Betty paid \$16,000 per acre.

35. $187 \div 8\frac{1}{2} = \frac{187}{1} \div \frac{17}{2}$

$= \frac{187}{1} \bullet \frac{2}{17}$

$= \frac{11 \bullet \cancel{17}}{1} \bullet \frac{2}{\cancel{17}} = 22$

Laura's car gets 22 miles per gallon.

37. $11 \div 2\frac{3}{4} = \frac{11}{1} \div \frac{11}{4}$

$= \frac{\cancel{11}}{1} \bullet \frac{4}{\cancel{11}} = 4$

Jill can make 4 batches of cookies.

39. $42 \div 4\frac{1}{5} = \frac{42}{1} \div \frac{21}{5} = \frac{42}{1} \bullet \frac{5}{21}$

$= \frac{2 \bullet \cancel{21}}{1} \bullet \frac{5}{\cancel{21}} = 10$

Maria earns \$10 per hour.

41. $440 \bullet 8\frac{3}{4} = \frac{440}{1} \bullet \frac{35}{4}$

$= \frac{\cancel{4} \bullet 110}{1} \bullet \frac{35}{\cancel{4}} = 3850$

Ryan paid \$3850 for the stock.

43. improper.

45. $2\frac{1}{3} \bullet 4\frac{2}{3} \bullet 3\frac{3}{8} = \frac{7}{3} \bullet \frac{14}{3} \bullet \frac{27}{8}$

$= \frac{7}{\cancel{3}} \bullet \frac{\cancel{2} \bullet 7}{\cancel{3}} \bullet \frac{\cancel{3} \bullet \cancel{3} \bullet 3}{\cancel{2} \bullet 4}$

$= \frac{147}{4} = 36\frac{3}{4}$

47. $4000 \bullet 2\frac{1}{2} = \frac{4000}{1} \bullet \frac{5}{2} =$

$= \frac{\cancel{2} \bullet 2000}{1} \bullet \frac{5}{\cancel{2}} = 10,000$

The Food Bank received \$10,000.

$10,000 \bullet 3\frac{1}{5} = \frac{10,000}{1} \bullet \frac{16}{5}$

$= \frac{\cancel{5} \bullet 2000}{1} \bullet \frac{16}{\cancel{5}} = 32,000$

United Fund received \$32,000.

49. 36, 129, 6183

51. 7180, 50

53. 56, 184, 6812, 70, 540

55. 70, 540

Problem Set 3.6

1.

$$\begin{array}{r} 3\frac{1}{4} \\ +5\frac{3}{14} \\ \hline 8\frac{4}{14} \end{array} = 8\frac{2}{7}$$

3.

$$\begin{array}{rcr} 3\frac{1}{3} \bullet \frac{5}{5} & = & 3\frac{5}{15} \\ 4\frac{1}{5} \bullet \frac{3}{3} & = & +4\frac{3}{15} \\ \hline & & 7\frac{8}{15} \end{array}$$

5.

$$\begin{array}{rcr} 5\frac{2}{3} \bullet \frac{3}{3} & = & 5\frac{6}{9} \\ +3\frac{5}{9} & = & +3\frac{5}{9} \\ \hline & & 8\frac{11}{9} \end{array}$$

$8\frac{11}{9} = 8 + 1\frac{2}{9} = 9\frac{2}{9}$

Problem Set 3.6

7.
$$7\frac{5}{9}\cdot\frac{4}{4} = 7\frac{20}{36}$$
$$+4\frac{5}{12}\cdot\frac{3}{3} = +4\frac{15}{36}$$
$$11\frac{35}{36}$$

9.
$$18\frac{2}{3}\cdot\frac{3}{3} = 18\frac{6}{9}$$
$$+14\frac{4}{9} = +14\frac{4}{9}$$
$$32\frac{10}{9}$$
$$32\frac{10}{9} = 32+1\frac{1}{9} = 33\frac{1}{9}$$

11.
$$5\frac{1}{5}\cdot\frac{8}{8} = 5\frac{8}{40}$$
$$2\frac{3}{8}\cdot\frac{5}{5} = 2\frac{15}{40}$$
$$+7\frac{3}{4}\cdot\frac{10}{10} = +7\frac{30}{40}$$
$$14\frac{53}{40}$$
$$14\frac{53}{40} = 14+1\frac{13}{40} = 15\frac{13}{40}$$

13.
$$9\frac{7}{8}$$
$$-6\frac{3}{8}$$
$$3\frac{4}{8} = 3\frac{1}{2}$$

15.
$$5\frac{3}{4}\cdot\frac{2}{2} = 5\frac{6}{8}$$
$$-2\frac{3}{8} = -2\frac{3}{8}$$
$$3\frac{3}{8}$$

17.
$$8\frac{4}{5}\cdot\frac{3}{3} = 8\frac{12}{15}$$
$$-2\frac{2}{3}\cdot\frac{5}{5} = -2\frac{10}{15}$$
$$6\frac{2}{15}$$

19.
$$9 = 8\frac{6}{6}$$
$$-2\frac{1}{6} = -2\frac{1}{6}$$
$$6\frac{5}{6}$$

21.
$$8\frac{1}{3}\cdot\frac{4}{4} = 8\frac{4}{12} = 7\frac{16}{12}$$
$$-2\frac{3}{4}\cdot\frac{3}{3} = -2\frac{9}{12} = -2\frac{9}{12}$$
$$5\frac{7}{12}$$

23.
$$26\frac{3}{8} = 26\frac{3}{8} = 25\frac{11}{8}$$
$$-11\frac{3}{4}\cdot\frac{2}{2} = -11\frac{6}{8} = -11\frac{6}{8}$$
$$14\frac{5}{8}$$

25.
$$74 = 73\frac{4}{4}$$
$$-\frac{3}{4} = -\frac{3}{4}$$
$$73\frac{1}{4}$$

27.
$$32\frac{5}{6}\cdot\frac{3}{3} = 32\frac{15}{18}$$
$$-24\frac{2}{9}\cdot\frac{2}{2} = -24\frac{4}{18}$$
$$8\frac{11}{18}$$

29.
$$2\frac{1}{4}\cdot\frac{5}{5} = 2\frac{5}{20}$$
$$+3\frac{1}{5}\cdot\frac{4}{4} = +3\frac{4}{20}$$
$$5\frac{9}{20}$$

Scott bought $5\frac{9}{20}$ pounds of candy.

31.
$$51\frac{1}{3}\cdot\frac{8}{8} = 51\frac{8}{24}$$
$$-46\frac{1}{8}\cdot\frac{3}{3} = -46\frac{3}{24}$$
$$5\frac{5}{24}$$

Jason is $5\frac{5}{24}$ inches taller than Kevin.

Problem Set 3.6

33.
$$27\frac{2}{5} \bullet \frac{2}{2} = 27\frac{4}{10}$$
$$+18\frac{3}{10} = +18\frac{3}{10}$$
$$45\frac{7}{10}$$

The total length is $45\frac{7}{10}$ centimeters.

35.
$$2\frac{1}{2} \bullet \frac{4}{4} = 2\frac{4}{8} = 1\frac{12}{8}$$
$$-1\frac{5}{8} = -1\frac{5}{8} = -1\frac{5}{8}$$
$$\frac{7}{8}$$

There is $\frac{7}{8}$ cup more orange juice than pineapple juice.

37.
$$2\frac{1}{2} \bullet \frac{3}{3} = 2\frac{3}{6}$$
$$+1\frac{1}{3} \bullet \frac{2}{2} = +1\frac{2}{6}$$
$$3\frac{5}{6}$$

Mary used $3\frac{5}{6}$ gallons of paint.

39.
$$40 = 39\frac{3}{3}$$
$$-27\frac{1}{3} = -27\frac{1}{3}$$
$$12\frac{2}{3}$$

It uses $12\frac{2}{3}$ gallons more water.

41.
$$6\frac{1}{2} \bullet \frac{5}{5} = 6\frac{5}{10}$$
$$+182\frac{3}{10} = +182\frac{3}{10}$$
$$188\frac{8}{10}$$
$$188\frac{8}{10} = 188\frac{4}{5}$$

Joel is $188\frac{4}{5}$ cm tall.

43.
$$53\frac{1}{3} \bullet \frac{4}{4} = 53\frac{4}{12}$$
$$53\frac{1}{3} \bullet \frac{4}{4} = 53\frac{4}{12}$$
$$43\frac{1}{4} \bullet \frac{3}{3} = 43\frac{3}{12}$$
$$+43\frac{1}{4} \bullet \frac{3}{3} = +43\frac{3}{12}$$
$$192\frac{14}{12}$$
$$192\frac{14}{12} = 192 + 1\frac{1}{6} = 193\frac{1}{6}$$

The distance is $193\frac{1}{6}$ cm.

45.
$$7\frac{1}{2} \bullet \frac{2}{2} = 7\frac{2}{4} = 6\frac{6}{4}$$
$$-2\frac{3}{4} = -2\frac{3}{4} = -2\frac{3}{4}$$
$$4\frac{3}{4}$$
$$15\frac{1}{2} \bullet \frac{2}{2} = 15\frac{2}{4} = 14\frac{6}{4}$$
$$-\ 4\frac{3}{4} = -\ 4\frac{3}{4} = -\ 4\frac{3}{4}$$
$$10\frac{3}{4}$$

Larry should freeze $10\frac{3}{4}$ more quarts of strawberries.

47.
$$4\frac{1}{4} \bullet 3 = \frac{17}{4} \bullet \frac{3}{1} = \frac{51}{4} = 12\frac{3}{4}$$
$$3\frac{1}{2} \bullet 4 = \frac{7}{2} \bullet \frac{4}{1}$$
$$= \frac{7}{\cancel{2}} \bullet \frac{\cancel{2} \bullet 2}{1} = 14$$
$$2\frac{3}{8} \bullet 5 = \frac{19}{8} \bullet \frac{5}{1} = \frac{95}{8} = 11\frac{7}{8}$$
$$12\frac{3}{4} \bullet \frac{2}{2} = 12\frac{6}{8}$$
$$14 = 14$$
$$+11\frac{7}{8} = +11\frac{7}{8}$$
$$37\frac{13}{8}$$
$$37\frac{13}{8} = 37 + 1\frac{5}{8} = 38\frac{5}{8}$$

The dressmaker needs $38\frac{5}{8}$ yards of fabric.

49. $51 > 32$

Problem Set 3.6

51. $0 < 24$

53. $$\begin{aligned} 7+36 \div 6 &= 7+6 \\ &= 13 \end{aligned}$$

55. $$\begin{aligned} 63 \div (5+4) &= 63 \div 9 \\ &= 7 \end{aligned}$$

57. $$\begin{aligned} &25+[8-(6-4)] = \\ &25+[8-2] = \\ &25+6 = 31 \end{aligned}$$

Problem Set 3.7

1. $\frac{4}{5} > \frac{3}{5}$

3. $\frac{1}{2} > \frac{3}{10}$, since $\frac{5}{10} > \frac{3}{10}$

5. $\frac{1}{4} < \frac{1}{3}$, since $\frac{3}{12} < \frac{4}{12}$

7. $\frac{7}{2} > \frac{5}{3}$, since $\frac{21}{6} > \frac{10}{6}$

9. $\frac{14}{15} > \frac{4}{5}$, since $\frac{14}{15} > \frac{12}{15}$

11. $\frac{7}{9} > \frac{5}{7}$, since $\frac{49}{63} > \frac{45}{63}$

13. $\frac{5}{12} < \frac{7}{11}$, since $\frac{55}{132} < \frac{84}{132}$

15. $$\begin{aligned} \frac{2}{3}+\left(\frac{5}{6}-\frac{5}{12}\right) &= \frac{2}{3}+\left(\frac{10}{12}-\frac{5}{12}\right) \\ &= \frac{2}{3}+\frac{5}{12} \\ &= \frac{8}{12}+\frac{5}{12} \\ &= \frac{13}{12} = 1\frac{1}{12} \end{aligned}$$

17. $$\begin{aligned} \frac{11}{2}-\left(\frac{3}{4}-\frac{1}{3}\right) &= \frac{11}{2}-\left(\frac{9}{12}-\frac{4}{12}\right) \\ &= \frac{11}{2}-\frac{5}{12} \\ &= \frac{66}{12}-\frac{5}{12} \\ &= \frac{61}{12} = 5\frac{1}{12} \end{aligned}$$

19. $$\begin{aligned} \frac{5}{6}+\frac{3}{8} \div \frac{1}{4} &= \frac{5}{6}+\frac{3}{8} \bullet \frac{4}{1} \\ &= \frac{5}{6}+\frac{3}{\cancel{4} \bullet 2} \bullet \frac{\cancel{4}}{1} \\ &= \frac{5}{6}+\frac{3}{2} \\ &= \frac{5}{6}+\frac{9}{6} \\ &= \frac{14}{6} = 2\frac{1}{3} \end{aligned}$$

21. $$\begin{aligned} 6 \bullet \frac{2}{5}-\frac{5}{4} &= \frac{6}{1} \bullet \frac{2}{5}-\frac{5}{4} \\ &= \frac{12}{5}-\frac{5}{4} \\ &= \frac{48}{20}-\frac{25}{20} \\ &= \frac{23}{20} = 1\frac{3}{20} \end{aligned}$$

23. $$\begin{aligned} \left(\frac{12}{5}-\frac{2}{3}\right)-\frac{1}{4} &= \left(\frac{36}{15}-\frac{10}{15}\right)-\frac{1}{4} \\ &= \frac{26}{15}-\frac{1}{4} \\ &= \frac{104}{60}-\frac{15}{60} \\ &= \frac{89}{60} = 1\frac{29}{60} \end{aligned}$$

25. $$\begin{aligned} \frac{9}{4}-2 \bullet \frac{9}{10} &= \frac{9}{4}-\frac{2}{1} \bullet \frac{9}{10} \\ &= \frac{9}{4}-\frac{18}{10} \\ &= \frac{45}{20}-\frac{36}{20} \\ &= \frac{9}{20} \end{aligned}$$

27. $$\begin{aligned} 9-\frac{4}{3}-\frac{5}{12} &= \frac{9}{1}-\frac{4}{3}-\frac{5}{12} \\ &= \frac{108}{12}-\frac{16}{12}-\frac{5}{12} \\ &= \frac{87}{12} = 7\frac{1}{4} \end{aligned}$$

29. $$\begin{aligned} \frac{9}{8} \div \frac{3}{2} \div \frac{4}{9} &= \frac{9}{8} \bullet \frac{2}{3} \div \frac{4}{9} \\ &= \frac{\cancel{3} \bullet 3}{\cancel{2} \bullet 4} \bullet \frac{\cancel{2}}{\cancel{3}} \div \frac{4}{9} \\ &= \frac{3}{4} \bullet \frac{9}{4} \\ &= \frac{27}{16} = 1\frac{11}{16} \end{aligned}$$

Problem Set 3.7

31. $9 \div \frac{3}{7} \div \frac{21}{2} = \frac{9}{1} \bullet \frac{7}{3} \div \frac{21}{2}$
$= \frac{3 \bullet \cancel{3}}{1} \bullet \frac{7}{\cancel{3}} \div \frac{21}{2}$
$= \frac{2\cancel{1}}{1} \bullet \frac{2}{2\cancel{1}}$
$= 2$

33. $\frac{2}{7} \div 6 \bullet \frac{23}{5} = \frac{2}{7} \bullet \frac{1}{6} \bullet \frac{23}{5}$
$= \frac{\cancel{2}}{7} \bullet \frac{1}{\cancel{2} \bullet 3} \bullet \frac{23}{5}$
$= \frac{1}{21} \bullet \frac{23}{5}$
$= \frac{23}{105}$

35. $\frac{3}{4} < \frac{7}{8}$, since $\frac{6}{8} < \frac{7}{8}$
Dana has enough fabric.

37. larger

39. $\frac{5}{12}, \frac{1}{3}, \frac{2}{9}$
since $\frac{15}{36} > \frac{12}{36} > \frac{8}{36}$

41. $\left(\frac{4}{3} + \frac{5}{7}\right) - \frac{4}{5} \bullet 2$
$= \left(\frac{28}{21} + \frac{15}{21}\right) - \frac{4}{5} \bullet 2$
$= \frac{43}{21} - \frac{4}{5} \bullet \frac{2}{1}$
$= \frac{43}{21} - \frac{8}{5}$
$= \frac{215}{105} - \frac{168}{105}$
$= \frac{47}{105}$

43. ones

45. Seven thousand, eight hundred two

47. Ninety thousand, eight

49. 4005

51. 4,008,000

Chapter 3 Additional Exercises

1. $\frac{4}{9} + \frac{2}{9} = \frac{6}{9} = \frac{2}{3}$

3. $\frac{17}{7} - \frac{3}{7} = \frac{14}{7} = 2$

5. $\frac{5}{12} + \frac{11}{12} = \frac{16}{12} = \frac{4}{3}$

7. $\frac{21}{10} - \frac{9}{10} = \frac{12}{10} = \frac{6}{5}$

9. $\frac{9}{21} + \frac{17}{21} + \frac{7}{21} = \frac{33}{21} = \frac{11}{7}$

11. $16 = 2 \bullet 2 \bullet 2 \bullet 2$
$32 = 2 \bullet 2 \bullet 2 \bullet 2 \bullet 2$
$\text{LCD} = 2 \bullet 2 \bullet 2 \bullet 2 \bullet 2 = 32$

13. $9 = 3 \bullet 3$
$10 = 2 \bullet 5$
$\text{LCD} = 2 \bullet 3 \bullet 3 \bullet 5 = 90$

15. $16 = 2 \bullet 2 \bullet 2 \bullet 2$
$36 = 2 \bullet 2 \bullet 3 \bullet 3$
$\text{LCD} = 2 \bullet 2 \bullet 2 \bullet 2 \bullet 3 \bullet 3$
$= 144$

17. $32 = 2 \bullet 2 \bullet 2 \bullet 2 \bullet 2$
$64 = 2 \bullet 2 \bullet 2 \bullet 2 \bullet 2 \bullet 2$
$\text{LCD} = 2 \bullet 2 \bullet 2 \bullet 2 \bullet 2 \bullet 2$
$= 64$

19. $20 = 2 \bullet 2 \bullet 5$
$35 = 5 \bullet 7$
$16 = 2 \bullet 2 \bullet 2 \bullet 2$
$\text{LCD} = 2 \bullet 2 \bullet 2 \bullet 2 \bullet 5 \bullet 7$
$= 560$

21. $\frac{2}{7} + \frac{3}{8} = \frac{2}{7} \bullet \frac{8}{8} + \frac{3}{8} \bullet \frac{7}{7}$
$= \frac{15}{56} + \frac{21}{56} = \frac{37}{56}$

23. $\frac{11}{8} - \frac{7}{18} = \frac{11}{8} \bullet \frac{9}{9} - \frac{7}{18} \bullet \frac{4}{4}$
$= \frac{99}{72} - \frac{28}{72} = \frac{71}{72}$

25. $\frac{5}{16} + \frac{7}{18} + \frac{3}{28} = \frac{5}{16} \bullet \frac{63}{63}$
$+ \frac{7}{18} \bullet \frac{56}{56} + \frac{3}{28} \bullet \frac{36}{36}$
$= \frac{315}{1008} + \frac{392}{1008} + \frac{108}{1008}$
$= \frac{815}{1008}$

Chapter 3 Additional Exercises

27. $\frac{29}{35}-\frac{7}{15} = \frac{29}{35}\bullet\frac{3}{3}-\frac{7}{15}\bullet\frac{7}{7}$
$= \frac{87}{105}-\frac{49}{105} = \frac{38}{105}$

29. $\frac{3}{4}+\frac{7}{8} = \frac{3}{4}\bullet\frac{2}{2}+\frac{7}{8}$
$= \frac{6}{8}+\frac{7}{8} = \frac{13}{8}$
There are $\frac{13}{8}$ cups of dry ingredients.

31. $\frac{1}{4}+\frac{7}{8}+\frac{3}{16}+\frac{1}{16}$
$= \frac{1}{4}\bullet\frac{4}{4}+\frac{7}{8}\bullet\frac{2}{2}+\frac{3}{16}+\frac{1}{16}$
$= \frac{4}{16}+\frac{14}{16}+\frac{3}{16}+\frac{1}{16}$
$= \frac{22}{16}$
$2-\frac{22}{16} = \frac{32}{16}-\frac{22}{16} = \frac{10}{16} = \frac{5}{8}$
There is $\frac{5}{8}$ inch of space left.

33. $7\frac{4}{9} = \frac{(9\times 7)+4}{9} = \frac{67}{9}$

35. $72\frac{3}{5} = \frac{(5\times 72)+3}{5} = \frac{363}{5}$

37. $5\overline{)38}$ = 7, $38-35=3$; $7\frac{3}{5}$

39. $10\overline{)525}$ = 52, $52-50=2$... $525-500=25$, $25-20=5$; $52\frac{1}{2}$

41. $6\frac{2}{3}\bullet 9 = \frac{20}{3}\bullet\frac{9}{1} = \frac{20}{\not{3}}\bullet\frac{\not{3}\bullet 3}{1}$
$= \frac{60}{1} = 60$

43. $6\frac{1}{8}\div\frac{8}{9} = \frac{49}{8}\bullet\frac{9}{8} = \frac{441}{64} = 6\frac{57}{64}$

45. $7\frac{1}{6}\bullet\frac{3}{5} = \frac{43}{6}\bullet\frac{3}{5} = \frac{129}{30}$
$= 4\frac{9}{30} = 4\frac{3}{10}$

47. $15\div 7\frac{1}{7} = \frac{15}{1}\div\frac{50}{7} = \frac{15}{1}\bullet\frac{7}{50}$
$= \frac{3\bullet\not{5}}{1}\bullet\frac{7}{\not{5}\bullet 10}$
$= \frac{21}{10} = 2\frac{1}{10}$

49. $25\frac{1}{2}\div 3 = \frac{51}{2}\div\frac{3}{1} = \frac{51}{2}\bullet\frac{1}{3}$
$= \frac{\not{3}\bullet 17}{2}\bullet\frac{1}{\not{3}} = \frac{17}{2} = 8\frac{1}{2}$
Each son receives $8\frac{1}{2}$ acres of land.

51. $9\frac{3}{8}\bullet\frac{3}{3} = 9\frac{9}{24}$
$+8\frac{7}{12}\bullet\frac{2}{2} = +8\frac{14}{24}$
$17\frac{23}{24}$

53. $16\frac{7}{9}\bullet\frac{2}{2} = 16\frac{14}{18}$
$-9\frac{5}{18} = -9\frac{5}{18}$
$7\frac{9}{18} = 7\frac{1}{2}$

55. $16 = 15\frac{8}{8}$
$-\frac{7}{8} = -\frac{7}{8}$
$15\frac{1}{8}$

57. $10\frac{3}{11}\bullet\frac{2}{2} = 10\frac{6}{22}$
$+4\frac{5}{22} = -4\frac{5}{22}$
$14\frac{11}{22} = 14\frac{1}{2}$

59. $17 = 16\frac{9}{9}$
$-14\frac{2}{9} = -14\frac{2}{9}$
$2\frac{7}{9}$

Chapter 3 Additional Exercises

61. $$\begin{array}{rclcl} 16\frac{1}{2}\cdot\frac{4}{4} & = & 16\frac{4}{8} & = & 15\frac{12}{8} \\ -12\frac{7}{8} & = & -12\frac{7}{8} & = & -12\frac{7}{8} \\ \hline & & & & 3\frac{5}{8} \end{array}$$

The first pole is $3\frac{5}{8}$ feet longer than the second pole.

63. $$49\frac{1}{2}\cdot\frac{1}{2} = \frac{99}{2}\cdot\frac{1}{2} = \frac{99}{4} = 24\frac{3}{4}$$

$$\begin{array}{rcl} 24\frac{3}{4}\cdot\frac{3}{3} & = & 24\frac{9}{12} \\ +23\frac{5}{6}\cdot\frac{2}{2} & = & +23\frac{10}{12} \\ \hline & & 47\frac{19}{12} = 48\frac{7}{12} \end{array}$$

It takes Joe $48\frac{7}{12}$ minutes to vacuum and dust.

65. $\frac{9}{8} > \frac{11}{10}$, since $\frac{45}{40} > \frac{44}{40}$

67. $\frac{7}{8} < \frac{15}{16}$, since $\frac{14}{16} > \frac{15}{16}$

69. $$\begin{aligned} 5+7\cdot\frac{3}{14} &= \frac{5}{1}+\frac{\cancel{7}}{1}\cdot\frac{3}{2\cdot\cancel{7}} \\ &= \frac{5}{1}+\frac{3}{2} \\ &= \frac{5}{1}\cdot\frac{2}{2}+\frac{3}{2} \\ &= \frac{10}{2}+\frac{3}{2} \\ &= \frac{13}{2} = 6\frac{1}{2} \end{aligned}$$

71. $$\begin{aligned} \frac{17}{16}-\frac{3}{8}-\frac{2}{5} &= \frac{85}{80}-\frac{30}{80}-\frac{32}{80} \\ &= \frac{23}{80} \end{aligned}$$

Chapter 3 Practice Test

1. $$\frac{7}{10}+\frac{9}{10} = \frac{16}{10} = \frac{8}{5}$$

3. $$\begin{aligned} \frac{5}{6}+\frac{7}{12} &= \frac{5}{6}\cdot\frac{2}{2}+\frac{7}{12} \\ &= \frac{10}{12}+\frac{7}{12} = \frac{17}{12} \end{aligned}$$

5. $$\begin{aligned} \frac{7}{6}-\frac{3}{8} &= \frac{7}{6}\cdot\frac{4}{4}-\frac{3}{8}\cdot\frac{3}{3} \\ &= \frac{28}{24}-\frac{9}{24} = \frac{19}{24} \end{aligned}$$

7. $$7\frac{5}{8} = \frac{(8\times 7)+5}{8} = \frac{61}{8}$$

9. $$\begin{array}{r} 3 \\ 5\overline{)17} \\ \underline{15} \\ 2 \end{array} \qquad 3\frac{2}{5}$$

11. $$\begin{array}{r} 803 \\ 9\overline{)7235} \\ \underline{72} \\ 035 \\ \underline{27} \\ 8 \end{array} \qquad 803\frac{8}{9}$$

13. $$\begin{aligned} 6\cdot 3\frac{1}{10} &= \frac{6}{1}\cdot\frac{31}{10} \\ &= \frac{\cancel{2}\cdot 3}{1}\cdot\frac{31}{\cancel{2}\cdot 5} \\ &= \frac{93}{5} = 18\frac{3}{5} \end{aligned}$$

15. $$\begin{aligned} 8\frac{1}{3}\div\frac{2}{9} &= \frac{25}{3}\div\frac{2}{9} \\ &= \frac{25}{3}\cdot\frac{9}{2} = \frac{25}{\cancel{3}}\cdot\frac{\cancel{3}\cdot 3}{2} \\ &= \frac{75}{2} = 37\frac{1}{2} \end{aligned}$$

17. $$\begin{aligned} 3\frac{1}{8}\div 4\frac{1}{4} &= \frac{25}{8}\div\frac{17}{4} = \frac{25}{8}\cdot\frac{4}{17} \\ &= \frac{25}{2\cdot\cancel{4}}\cdot\frac{\cancel{4}}{17} = \frac{25}{34} \end{aligned}$$

19. $$\begin{array}{rcl} 3\frac{1}{4}\cdot\frac{5}{5} & = & 3\frac{5}{20} \\ 2\frac{2}{5}\cdot\frac{4}{4} & = & 2\frac{8}{20} \\ +6\frac{3}{10}\cdot\frac{2}{2} & = & +6\frac{6}{20} \\ \hline & & 11\frac{19}{20} \end{array}$$

21. $$\begin{array}{rcl} 9 & = & 8\frac{5}{5} \\ -5\frac{2}{5} & = & -5\frac{2}{5} \\ \hline & & 3\frac{3}{5} \end{array}$$

23. $\frac{7}{8} < \frac{8}{9}$, since $\frac{63}{72} < \frac{64}{72}$

25. $\frac{4}{5} + \frac{7}{6} - \frac{2}{3} = \frac{24}{30} + \frac{35}{30} - \frac{20}{30}$
$= \frac{39}{30} = 1\frac{9}{30} = 1\frac{3}{10}$

27. $2\frac{1}{2} \bullet 584 = \frac{5}{2} \bullet \frac{584}{1}$
$= \frac{5}{\not 2} \bullet \frac{\not 2 \bullet 292}{1} = 1460$
Nigel spends \$1460 for gasoline.

29. $36\frac{7}{10} = 36\frac{7}{10}$
$+25\frac{4}{5} \bullet \frac{2}{2} = +25\frac{8}{10}$
$61\frac{15}{10}$
$61\frac{15}{10} = 61 + 1\frac{1}{2} = 62\frac{1}{2}$
The total length of the pieces of wire is $62\frac{1}{2}$ cm.

Problem Set 4.1

1. hundredths

3. thousandths

5. Sixteen and three tenths

7. Three and nine hundred fifty-four thousandths

9. Eight hundred fifty-two thousandths

11. Six and nine hundredths

13. Seven hundred nine and six tenths

15. Five thousandths

17. Seventy-two and six hundred ninety-three thousand, four hundred fifty-seven millionths

19. Sixty thousand, thirty and eighty-four thousandths

21. Fifty-one dollars and thirty-two cents

23. Two dollars and nine cents

25. Eight hundred thirty-two dollars and five cents

27. Five and two hundredths dollars

29. Twenty and eighty-five hundredths dollars

31. Twelve thousand, three hundred five and eight hundredths dollars

33. 75.44

35. 8.70252

37. 40.06

39. 7005.039

41. 10,003.32

43. 406.000305

45. \$43.05

47. \$840.12

49. $9\frac{73}{100}$

51. $\frac{9}{10,000}$

53. $765\frac{8}{10} = 765\frac{4}{5}$

55. $43\frac{125}{1000} = 43\frac{1}{8}$

57. Fifty-seven hundred thousandths

59. right

61. and

63. 2.231

65. 0.000009

67. $3842 \approx 3840$

Problem Set 4.1

69. $1520 \approx 1500$

71. $9971 \approx 10{,}000$

73. $$\frac{11}{3} \div \frac{33}{1} = \frac{11}{3} \bullet \frac{1}{33} = \frac{\cancel{11}}{3} \bullet \frac{1}{3 \bullet \cancel{11}} = \frac{1}{9}$$

75. $$\frac{21}{10} \div \frac{14}{15} = \frac{21}{10} \bullet \frac{15}{14} = \frac{3 \bullet \cancel{7}}{2 \bullet \cancel{5}} \bullet \frac{3 \bullet \cancel{5}}{2 \bullet \cancel{7}} = \frac{9}{4} = 2\frac{1}{4}$$

Problem Set 4.2

1. $7.35 > 7.3$

3. $0.04 > 0.007$

5. $62.82 < 62.89$

7. $4 > 3.99$

9. $0.007 < 0.06$

11. $7.006 < 7.6$

13. $0.8 > 0.09$

15. $5.7834 < 5.78345$

17. $0.25 < 2.5$

19. $6.32 \approx 6.3$

21. $5.06 \approx 5.1$

23. $0.385 \approx 0.4$

25. $19.3579 \approx 19.4$

27. $0.445 \approx 0.4$

29. $3.51948 \approx 3.5$

31. $5.139 \approx 5.14$

33. $0.918 \approx 0.92$

35. $6.722 \approx 6.72$

37. $9.004 \approx 9.00$

39. $9.1345 \approx 9.13$

41. $0.12687 \approx 0.13$

43. $0.2374 \approx 0.237$

45. $6.8095 \approx 6.810$

47. $6.0008 \approx 6.001$

49. $8.1032 \approx 8.103$

51. $0.03824 \approx 0.038$

53. $9.99999 \approx 10.000$

55. $4382.1745 \approx 4382.2$

57. $4382.1745 \approx 4000$

59. $4382.1745 \approx 4400$

61. $893.1678 \approx 893.17$

63. $893.1678 \approx 893.168$

65. $893.1678 \approx 900$

67. $0.1854 \approx 0.19$ m

69. $1.60935 \approx 1.61$ km

71. 0.58, 0.52, 0.51, 0.508

73. $$\frac{15}{19} - \frac{8}{19} = \frac{7}{19}$$

75. $$\frac{9}{15} + \frac{21}{15} = \frac{30}{15} = 2$$

77. $$\begin{aligned}\frac{3}{4} + \frac{5}{9} \div \frac{2}{3} &= \frac{3}{4} + \frac{5}{9} \bullet \frac{3}{2} \\ &= \frac{3}{4} + \frac{5}{\cancel{3} \bullet 3} \bullet \frac{\cancel{3}}{2} \\ &= \frac{3}{4} + \frac{5}{6} \\ &= \frac{3}{4} \bullet \frac{3}{3} + \frac{5}{6} \bullet \frac{2}{2} \\ &= \frac{9}{12} + \frac{10}{12} \\ &= \frac{19}{12} \\ &= 1\frac{7}{12}\end{aligned}$$

Problem Set 4.2

79. $\frac{3}{5} \bullet 25 \div \frac{1}{3} = \frac{3}{5} \bullet \frac{25}{1} \div \frac{1}{3}$
$= \frac{3}{\not{5}} \bullet \frac{\not{5} \bullet 5}{1} \div \frac{1}{3}$
$= \frac{15}{1} \bullet \frac{3}{1}$
$= 45$

81. $\frac{11}{3} - \frac{3}{4} \div \frac{7}{8} = \frac{11}{3} - \frac{3}{4} \bullet \frac{8}{7}$
$= \frac{11}{3} - \frac{3}{\not{4}} \bullet \frac{2 \bullet \not{4}}{7}$
$= \frac{11}{3} - \frac{6}{7}$
$= \frac{11}{3} \bullet \frac{7}{7} - \frac{6}{7} \bullet \frac{3}{3}$
$= \frac{77}{21} - \frac{18}{21}$
$= \frac{59}{21}$
$= 2\frac{17}{21}$

Problem Set 4.3

1. 0.657 + 0.34 = 0.997

3. 7.834 + 2.3 = 10.134

5. 2.004 + 184.9 = 186.904

7. 50.3864 + 70.0295 = 120.4159

9. 6.347 + 2.98 + 5.03 = 14.357

11. 122.48 + 17.652 + 10.091 = 150.223

13. 204. + 5.19 + 0.6 = 209.79

15. 4.8 + 16.347 + 321.098 + 18.13 = 360.375

17. 19.18 + 7.314 + 5613.2048 + 8.132 = 5647.8308

19. 7784.3 + 205.912 + 8613.09 + 3.5178 = 16606.8198

21.

1.99	≈	2
19.95	≈	20
+274.98	≈	+275
		297

The cost is about $297.00.

23.

72.34	≈	72
4.8	≈	5
+36.75	≈	+37
		114

25.

6.59	≈	7
0.89	≈	1
+12.435	≈	+12
		20

27.

6.845	≈	6.8
+13.6432	≈	+13.6
		20.4

29.

765.054	≈	770
+ 9.87	≈	+ 10
		780

31. 327.8 + 417.5 + 84.3 + 495.6 = 1325.2

Ken drove 1325.2 miles.

Problem Set 4.3

33.
$$\begin{array}{r} 6.27 \\ 4.38 \\ 4.97 \\ 3.18 \\ +\underline{2.79} \\ 21.59 \end{array}$$
The total rainfall was 21.59 inches.

35.
$$\begin{array}{r} 91.6 \\ +\underline{10.7} \\ 102.3 \end{array}$$
The gasoline cost 102.3 cents per gallon.

37.
$$\begin{array}{r} 54.95 \\ 21.38 \\ 6.79 \\ +\underline{74.65} \\ 157.77 \end{array}$$
Karen's bill was $157.77.

39.
$$\begin{array}{r} 9.01 \\ +\underline{2731.08} \\ 2740.09 \end{array}$$
The new average was 2740.09.

41.
$$\begin{array}{r} 0.89 \\ 5.97 \\ 27.99 \\ +\underline{39.88} \\ 74.73 \end{array}$$
Jeff owes the store $74.73.

43.
$$\begin{array}{r} 29.4 \\ +\underline{39.8} \\ 69.2 \end{array}$$
Ramona eats 69.2 grams of cereal.

45. right

47.
$$\begin{array}{r} 80{,}978.34 \\ 52{,}999.89 \\ 35.04 \\ 7{,}608.9 \\ +\underline{23{,}321.99} \\ 164{,}944.16 \end{array}$$

49.
$$\begin{array}{rcr} 7.98 & \approx & 8 \\ 34.25 & \approx & 34 \\ 13.95 & \approx & 14 \\ 57.32 & \approx & 57 \\ 238.64 & \approx & 239 \\ 44.49 & \approx & 44 \\ 73.80 & \approx & 74 \\ +\underline{326.18} & \approx & +\underline{326} \\ & & 796 \end{array}$$
The bill is about $796.

51.
$16 = 2 \bullet 2 \bullet 2 \bullet 2$
$28 = 2 \bullet 2 \bullet 7$
$32 = 2 \bullet 2 \bullet 2 \bullet 2 \bullet 2$
$\text{LCD} = 2 \bullet 2 \bullet 2 \bullet 2 \bullet 2 \bullet 7$
$= 224$

53.
$6 = 2 \bullet 3$
$12 = 2 \bullet 2 \bullet 3$
$27 = 3 \bullet 3 \bullet 3$
$\text{LCD} = 2 \bullet 2 \bullet 3 \bullet 3 \bullet 3$
$= 108$

55.
$$\begin{aligned} \frac{3}{8} + \frac{1}{32} &= \frac{3}{8} \bullet \frac{4}{4} + \frac{1}{32} \\ &= \frac{12}{32} + \frac{1}{32} \\ &= \frac{13}{32} \end{aligned}$$

57.
$$\begin{aligned} \frac{7}{9} - \frac{2}{7} &= \frac{7}{9} \bullet \frac{7}{7} - \frac{2}{7} \bullet \frac{9}{9} \\ &= \frac{49}{63} - \frac{18}{63} \\ &= \frac{31}{63} \end{aligned}$$

59.
$$\begin{aligned} \frac{19}{14} - \frac{23}{35} &= \frac{19}{14} \bullet \frac{5}{5} - \frac{23}{35} \bullet \frac{2}{2} \\ &= \frac{95}{70} - \frac{46}{70} \\ &= \frac{49}{70} \\ &= \frac{7}{10} \end{aligned}$$

Problem Set 4.4

1.
$$\begin{array}{r} 5\ \ 18 \\ \not{6}\ .\ \not{8}\ 9 \\ -\underline{4\ .\ 9\ 2} \\ 1\ .\ 9\ 7 \end{array}$$

Problem Set 4.4

3.

```
   8 14
0 . 9 4   (9, 4 crossed out)
-0 . 6 8
0 . 2 6
```

5.

```
  15
5 5   12      (second 5 crossed out)
6 6 . 2       (6, 6, 2 crossed out)
-3 7 . 5
2 8 . 7
```

7.

```
  3   17
9 4 . 7 8 2   (4, 7 crossed out)
-  3 . 9 2 0
9 0 . 8 6 2
```

9.

```
        9
      8 10 10
8 7 . 9 0 0   (9, 0, 0 crossed out; 10 crossed out)
-  3 . 8 4 5
8 4 . 0 5 5
```

11.

```
    11
  4 1    10
7 5 2 . 0     (5, 2, 0 crossed out; 1 crossed out)
-  3 6 . 9
7 1 5 . 1
```

13.

```
             6 16
1 6 . 8 9 7 6    (7, 6 crossed out)
-  4 . 8 0 3 7
1 2 . 0 9 3 9
```

15.

```
        9
      7 10 10
9 . 0 8 0 0   (8, 0, 0 crossed out; 10 crossed out)
-3 . 0 0 4 2
6 . 0 7 5 8
```

17.

```
    9  9
2  10 10 10
3 . 0 0 0     (3, 0, 0, 0 crossed out; first two 10s crossed out)
-2 . 9 9 9
0 . 0 0 1
```

19.

```
     9  9 13
5  10 10 3 17
6 . 0 0 4 7   (6, 0, 0, 4, 7 crossed out; 10, 10, 3 crossed out)
-5 . 9 9 8 8
0 . 0 0 5 9
```

21.

```
   8 14
0 . 9 4   (9, 4 crossed out)
-0 . 8 6
0 . 0 8
```

23.

```
  13   11
8 3    1 10
9 4 . 2 0     (9, 4, 2, 0 crossed out; 3, 1 crossed out)
-3 7 . 8 5
5 6 . 3 5
```

25.

```
      12
8    2 14
9 . 3 4 5     (9, 3, 4 crossed out; 2 crossed out)
-7 . 8 6 0
1 . 4 8 5
```

27.

```
   3  10
8 4 . 0       (4, 0 crossed out)
-  3 . 8
8 0 . 2
```

29.

```
        9
     5 10 10
0 . 6 0 0     (6, 0, 0 crossed out; first 10 crossed out)
-0 . 3 1 9
0 . 2 8 1
```

31.

```
         9   9
   3  10 10 10
7 4 . 0 0 0   (4, 0, 0, 0 crossed out; first two 10s crossed out)
-  3 . 6 8 9
7 0 . 3 1 1
```

33.

```
4.719
-3.602
1.117
```

35.

```
           9
6 11 8   10 10
7 1 9 . 0 0   (7, 1, 9, 0, 0 crossed out; first 10 crossed out)
-  3 2 . 9 8
6 8 6 . 0 2
```

37.

```
  9     9  9  9
5 10   10 10 10 14
6 0 . 0 0 0 4     (all top digits crossed out; 10s crossed out)
-4 2 . 0 5 7 8
1 7 . 9 4 2 6
```

39.

```
  9  9    9  9
1 10 10  10 10 10
2 0 0 . 0 0 0     (all top digits crossed out; first four 10s crossed out)
-    0 . 3 7 8
1 9 9 . 6 2 2
```

41.

```
 103.25 ≈  103
- 79.95 ≈ - 80
            23
```

The difference in the cost is about $23.

43.

```
 6.34  ≈  6
-1.085 ≈ -1
          5
```

45.

```
 785.95 ≈  790
-450.85 ≈ -450
           340
```

Problem Set 4.4

47.
16
8 6 14
8 ~~0~~ . ~~7~~ ~~4~~
-34.95
54.79
Mika kept a part worth $54.79.

49.
9
4 ~~1~~0 10
~~5~~ . ~~0~~ ~~0~~
-3.76
1.24
Lisa received $1.24 in change.

51.
13 11
7 ~~3~~ ~~1~~ 15
~~8~~ ~~4~~ . ~~2~~ ~~5~~
-56.37
27.88
Carlos needs $27.88.

53.
9
7 ~~1~~0 16 7 11
4 ~~8~~ , ~~0~~ ~~6~~ ~~8~~ . ~~1~~
-47,894.2
173.9
Scott traveled 173.9 miles.

55.
9 13
3 ~~1~~0 ~~3~~ 10
~~4~~ ~~0~~ ~~4~~ . ~~0~~ 6
-394.56
9.50
Alicia received $9.50.

57.
5 12
1 2 ~~6~~ . ~~2~~
-114.6
11.6
The price rose 11.6 cents.

59.
12
7 ~~2~~ 14
2 6 ~~8~~ . ~~3~~ ~~4~~
-47.85
220.49
The value of the clothing Keith kept was $220.49.

61.
11
2 ~~1~~ 10
~~3~~ ~~2~~ . ~~0~~ 9
-23.87
8.22
The difference in length is 8.22 centimeters.

63.
12.324
+12.324
24.648

2 10 7 10
~~3~~ ~~0~~ . 6 ~~8~~ ~~0~~
-24.648
6.032
The length of x is 6.032 meters.

65.
74.062348
-55.098430
18.963918

67.
2009.36
1094.32
+ 437.84
3541.52

1379.89
+ 96.22
1476.11

3541.52
-1476.11
2065.41
Juan's checking account balance was $2065.41.

69.
742
× 30
22260

71. $8\frac{7}{9} = \frac{(9 \times 8) + 7}{9} = \frac{79}{9}$

73. $15\frac{3}{4} = \frac{(4 \times 15) + 3}{4} = \frac{63}{4}$

75.
40
8)327
32
07

$40\frac{7}{8}$

77.
92
9)832
81
22
18
4

$92\frac{4}{9}$

Problem Set 4.5

1.
6.3
× 5
31.5

3.
7.2
× 0.9
6.48

5.
0.8
× 0.7
0.56

7.
85
× 0.03
2.55

Problem Set 4.5

9.
$$\begin{array}{r} 74.5 \\ \times\underline{\ \ 2.1} \\ 7\ 45 \\ \underline{149\ 0\ \ } \\ 156.45 \end{array}$$

11.
$$\begin{array}{r} 76 \\ \times\underline{0.005} \\ 0.380 \end{array}$$

13.
$$\begin{array}{r} 0.832 \\ \times\underline{\ 0.41} \\ 832 \\ \underline{\ \ 3328\ } \\ 0.34112 \end{array}$$

15.
$$\begin{array}{r} 6.98 \\ \times\underline{\ \ \ \ \ 50} \\ 349.00 \end{array}$$

17.
$$\begin{array}{r} 89.98 \\ \times\underline{\ \ \ \ \ 10} \\ 899.80 \end{array}$$

19.
$$\begin{array}{r} 7.142 \\ \times\underline{\ \ \ 3.2} \\ 1\ 4284 \\ \underline{21\ 426\ \ } \\ 22.8544 \end{array}$$

21.
$$\begin{array}{r} 52.36 \\ \times\underline{\ \ 8.14} \\ 2\ 0944 \\ 5\ 236\ \\ \underline{418\ 88\ \ \ } \\ 426.2104 \end{array}$$

23.
$$\begin{array}{r} 74.32 \\ \times\underline{10.05} \\ 3\ 7160 \\ \underline{743\ 2\ \ \ \ } \\ 746.9160 \end{array}$$

25.
$$\begin{array}{r} 7.89 \\ \times\underline{\ \ \ \ 10} \\ 78.90 \end{array}$$

27.
$$\begin{array}{r} 0.51 \\ \times\underline{0.0034} \\ 204 \\ \underline{\ \ \ \ \ 153\ } \\ 0.001734 \end{array}$$

29.
$$\begin{array}{r} 9.63 \\ \times\underline{7.034} \\ 3852 \\ 2889\ \\ \underline{67\ 41\ \ \ \ } \\ 67.73742 \end{array}$$

31.
$$\begin{array}{r} 6.783 \\ \times\underline{\ \ \ \ \ \ 100} \\ 678.300 \end{array}$$

33.
$$\begin{array}{r} 94.3 \\ \times\underline{\ 0.01} \\ 0.943 \end{array}$$

35.
$$\begin{array}{r} 640.2 \\ \times\underline{\ \ \ 7.1} \\ 64\ 02 \\ \underline{4481\ 4\ \ } \\ 4545.42 \end{array}$$

37.
$$\begin{array}{r} 62 \\ \times\underline{0.001} \\ 0.062 \end{array}$$

39.
$$\begin{array}{r} 6.38 \\ \times\underline{2.005} \\ 3190 \\ \underline{12\ 76\ \ \ \ } \\ 12.79190 \end{array}$$

41.
$$\begin{array}{r} 5.382 \\ \times\underline{\ \ \ 0.04} \\ 0.21528 \end{array}$$

43.
$$\begin{array}{r} 650 \\ \times\underline{1.00007} \\ 4550 \\ \underline{650\ \ \ \ \ \ \ \ \ } \\ 650.04550 \end{array}$$

45. \$5.94 = 594 cents

47. \$78.23 = 7823 cents

49. \$0.25 = 25 cents

51. 8642 cents = \$86.42

53. 428 cents = \$4.28

55. 2 cents = \$0.02

57. 6 × \$4.69 = 6 × \$5 = \$30

59. 68.3 × 9 = 70 × 9 = 630

61. 6 × 11.95 = 6 × 12 = 72
The cost is about \$72.

63.
$$\begin{array}{r} 36.5 \\ \times\underline{8.32} \\ 730 \\ 10\ 95\ \\ \underline{292\ 0\ \ \ } \\ 303.680 \end{array} \qquad \begin{array}{r} 37 \\ \times\underline{\ \ 8} \\ 296 \end{array}$$

Problem Set 4.5

65.
$$\begin{array}{r} 540.31 \\ \times\ 70.58 \\ \hline 43\ 2248 \\ 270\ 155\ \ \ \\ 37\ 821\ 7\ \ \ \ \ \ \\ \hline 38{,}135.0798 \end{array} \qquad \begin{array}{r} 540 \\ \times\ 71 \\ \hline 540 \\ 37\ 80\ \ \\ \hline 38{,}340 \end{array}$$

67.
$$\begin{array}{r} 1.74 \\ \times\ 12 \\ \hline 3\ 48 \\ 17\ 4\ \ \\ \hline 20.88 \end{array}$$
Jean paid $20.88.

69.
$$\begin{array}{r} 42.5 \\ \times 37.2 \\ \hline 8\ 50 \\ 297\ 5\ \ \\ 1275\ \ \ \ \\ \hline 1581.00 \end{array}$$
The area is 1581 square meters.

71.
$$\begin{array}{r} 59.95 \\ \times\ 6 \\ \hline 359.70 \end{array}$$
Kathy paid $359.70 for the dresses.

73.
$$\begin{array}{r} 5.75 \\ \times\ 25 \\ \hline 28\ 75 \\ 115\ 0\ \ \\ \hline 143.75 \end{array} \qquad \begin{array}{r} 7.89 \\ \times\ 19 \\ \hline 71\ 01 \\ 78\ 9\ \ \\ \hline 149.91 \end{array} \qquad \begin{array}{r} 143.75 \\ +149.91 \\ \hline 293.66 \end{array}$$
Andrew earned $293.66.

75.
$$\begin{array}{r} 16.4 \\ \times 102.9 \\ \hline 14\ 76 \\ 32\ 8\ \ \\ 164\ \ \ \ \ \ \\ \hline 1687.56 \end{array}$$
The gas costs 1688 cents or $16.88.

77. sum

79. left

81.
$$\begin{array}{r} 724.3 \\ \times\ 0.679 \\ \hline 491.7997 \end{array}$$

83. $78 \times 0.00001 = 0.00078$

85.
$$\begin{aligned} 11\frac{1}{3} \bullet 4\frac{1}{5} &= \frac{34}{3} \bullet \frac{21}{5} \\ &= \frac{34}{\cancel{3}} \bullet \frac{\cancel{3} \bullet 7}{5} \\ &= \frac{238}{5} = 47\frac{3}{5} \end{aligned}$$

87.
$$\begin{aligned} 1\frac{5}{31} \bullet 93 &= \frac{36}{31} \bullet \frac{93}{1} \\ &= \frac{36}{\cancel{31}} \bullet \frac{3 \bullet \cancel{31}}{1} \\ &= 108 \end{aligned}$$

89.
$$\begin{aligned} 6\frac{3}{8} \div 17 &= \frac{51}{8} \div \frac{17}{1} \\ &= \frac{51}{8} \bullet \frac{1}{17} \\ &= \frac{3 \bullet \cancel{17}}{8} \bullet \frac{1}{\cancel{17}} \\ &= \frac{3}{8} \end{aligned}$$

91.
$$\begin{aligned} 9\frac{1}{6} \div 3\frac{1}{8} &= \frac{55}{6} \div \frac{25}{8} \\ &= \frac{55}{6} \bullet \frac{8}{25} \\ &= \frac{\cancel{5} \bullet 11}{\cancel{2} \bullet 3} \bullet \frac{\cancel{2} \bullet 4}{\cancel{5} \bullet 5} \\ &= \frac{44}{15} \\ &= 2\frac{14}{15} \end{aligned}$$

93.
$$\begin{aligned} 230 \div 11\frac{1}{2} &= \frac{230}{1} \div \frac{23}{2} \\ &= \frac{230}{1} \bullet \frac{2}{23} \\ &= \frac{10 \bullet \cancel{23}}{1} \bullet \frac{2}{\cancel{23}} \\ &= 20 \end{aligned}$$
The car can travel 20 miles on one gallon of gasoline.

Problem Set 4.6

1.
$$\begin{array}{r} 3.92 \\ 2\overline{)7.84} \\ \underline{6}\ \ \ \ \ \\ 1\ 8\ \ \\ \underline{1\ 8}\ \ \\ 04 \\ \underline{4} \\ 0 \end{array}$$

Problem Set 4.6

3.
```
     7.4
  6)44.4
    42
     2 4
     2 4
       0
```

5.
```
      0.666  ≈ 0.67
  24)16.000
     14 4
      1 60
      1 44
         160
         144
           6
```

7.
```
      3.06
  12)36.72
     36
      0 72
        72
         0
```

9.
```
     6.371  ≈ 6.37
  9)57.342
    54
     3 3
     2 7
       64
       63
        12
         9
         3
```

11.
```
    0.875  ≈ 0.88
  8)7.000
    6 4
      60
      56
       40
       40
        0
```

13.
```
             59.5
  0.04)2.38 0
       --'  --'
        2 0
          38
          36
           2 0
           2 0
             0
```

15.
```
           3.7
  4.2)15.5 4
      _'   _'
      12 6
        2 9 4
        2 9 4
            0
```

17.
```
      3.095  ≈ 3.10
  24)74.300
     72
      2 30
      2 16
         140
         120
          20
```

19.
```
             0.5
  0.56)0.28 0
       --'  --'
          28 0
             0
```

21.
```
     4.05
  6)24.30
    24
     0 30
       30
        0
```

23.
```
            6.75
  5.6)37.8 00
      _'   _'
      33 6
       4 2 0
       3 9 2
         2 80
         2 80
            0
```

25.
```
              123.529  ≈ 123.53
  0.34)42.00 000
       --'  --'
        34
         8 0
         6 8
         1 20
         1 02
            18 0
            17 0
             1 00
               68
               320
               306
                14
```

Problem Set 4.6

27.

```
      87.713  ≈ 87.71
  9)789.420
    72
     69
     63
      6 4
      6 3
        12
         9
         30
         27
          3
```

29.

```
       0.933  ≈ 0.93
  15)14.000
     13 5
        50
        45
         50
         45
          5
```

31.

```
       0.0051  ≈ 0.01
  42)0.2165
       210
        65
        42
        23
```

33.

```
             3 04.166  ≈ 304.17
  0.24)73.00 000
    --'    --'
        72
        1 00
          96
           4 0
           2 4
           1 60
           1 44
             160
             144
              16
```

35.

```
               265.555  ≈ 265.56
  0.018)4.780 000
     ---'   ---'
          3 6
          1 18
          1 08
            100
             90
             10 0
              9 0
                100
                 90
                 10
```

37. 74.26 ÷ 10 = 7.426

39. 2.53 ÷ 0.01 = 253

41. 5.38 ÷ 1000 = 0.00538

43. 76.3 ÷ 0.1 = 763

45. 55.2 ÷ 10.9 ≈ 60 ÷ 10 ≈ 6

47. 100.3 ÷ 4.2 ≈ 100 ÷ 4 ≈ 25

49. 14.75 ÷ 3.35 ≈ 15 ÷ 3 ≈ 5
Karen bought about 5 plants.

51.

```
      525.75
  6)3154.50
    30
     15
     12
      34
      30
       4 5
       4 2
         30
         30
          0
```

Linda's rent is \$525.75 per month.

53.

```
        6.35
  18)114.30
     108
       6 3
       5 4
         90
         90
          0
```

Each person paid \$6.35.

55.

```
          8 0.
  5.2)416.0
     -'   --'
      416
        0 0
```

Ken typed 80 words per minute.

57.

```
         642.35
  36)23124.60
     216
      152
      144
        84
        72
        12 6
        10 8
         1 80
         1 80
            0
```

Delores pays \$642.35 per month.

Problem Set 4.6

59. $0.69\overline{)5.38}$ → $69\overline{)538.00}$ = 7.79 ≈ 7.8

```
        7.79  ≈ 7.8
0.69)5.38 00
      4 83
        55 0
        48 3
         6 70
         6 21
           49
```

Joan bought 7.8 pounds of oranges.

61.

```
  18753.3            2 6.30
 -18274.6     18.2)478.7 00
    478.7          364
                   114 7
                   109 2
                     5 5 0
                     5 4 6
                         40
```

Jack got 26.3 miles per gallon.

63. $\dfrac{74+71+43}{3} = 62.67$

65. $\dfrac{19.3+16.4+17.8}{3} = 17.83$

67. $\dfrac{9.8+63.4+18.5}{3} = 30.57$

69. $\dfrac{85+79+96+90+88}{5} = 87.6$

Jill's average score is 87.6.

71. $\dfrac{30{,}000+24{,}500+54{,}000}{3}$

$= 36{,}166.67$

Their average salary was $36,166.67.

73. $\dfrac{8.1+6.4+3.2+5.7}{4} = 5.85$

Their average weight was 5.85 pounds

75. $\dfrac{10.7+15.2+14.3+16.4+9.8}{5}$

$= 13.28$

Jon bought an average of 13.28 gallons of gasoline.

77. dividend, divisor

79. left

81. average

83. 47,040 ÷ 84 = 560

The monthly payment is $560.

85. 400 × 3 = 1200

250 + 385 = 635

1200 - 635 = 565

Susan's dinner should have 565 calories.

87. $15\frac{3}{4}\cdot\frac{2}{2} = 15\frac{6}{8}$

$+\ 7\frac{5}{8} = +\ 7\frac{5}{8}$

$22\frac{11}{8} = 23\frac{3}{8}$

89. $11\frac{1}{8}\cdot\frac{3}{3} = 11\frac{3}{24} = 10\frac{27}{24}$

$-\ 3\frac{5}{6}\cdot\frac{4}{4} = -\ 3\frac{20}{24} = -\ 3\frac{20}{24}$

$7\frac{7}{24}$

91. $18\frac{3}{8} = 18\frac{3}{8} = 17\frac{11}{8}$

$-\ 6 = -\ 5\frac{8}{8} = -\ 5\frac{8}{8}$

$12\frac{3}{8}$

93. $27 = 26\frac{9}{9}$

$-\ 6\frac{7}{9} = -\ 6\frac{7}{9}$

$20\frac{2}{9}$

95. $54\frac{1}{6}\cdot\frac{2}{2} = 54\frac{2}{12} = 53\frac{14}{12}$

$-38\frac{5}{12} = -38\frac{5}{12} = -38\frac{5}{12}$

$15\frac{9}{12}$

$15\frac{9}{12} = 15\frac{3}{4}$

Problem Set 4.7

Problem Set 4.7

1.
$$\begin{array}{r} 0.4 \\ 5\overline{)2.0} \\ \underline{2\ 0} \\ 0 \end{array}$$

3.
$$\begin{array}{r} 0.15 \\ 20\overline{)3.00} \\ \underline{2\ 0\ \ } \\ 1\ 00 \\ \underline{1\ 00} \\ 0 \end{array}$$

5.
$$\begin{array}{r} 0.4545 \\ 11\overline{)5.0000} \\ \underline{4\ 4\ \ \ \ \ \ } \\ 60\ \ \ \ \\ \underline{55}\ \ \ \ \\ 50\ \ \ \\ \underline{44}\ \ \ \\ 60\ \\ \underline{55}\ \\ 6\ \end{array} = 0.\overline{45}$$

7.
$$\begin{array}{r} 0.9166 \\ 12\overline{)11.0000} \\ \underline{10\ 8\ \ \ \ \ \ } \\ 20\ \ \ \ \\ \underline{12}\ \ \ \ \\ 80\ \ \ \\ \underline{72}\ \ \ \\ 80\ \\ \underline{72}\ \\ 8\ \end{array} = 0.91\overline{6}$$

9. $2\frac{1}{2} = 2 + 0.5 = 2.5$
$$\begin{array}{r} 0.5 \\ 2\overline{)1.0} \\ \underline{1\ 0} \\ 0 \end{array}$$

11.
$$\begin{array}{r} 0.4848 \\ 33\overline{)16.0000} \\ \underline{13\ 2\ \ \ \ \ \ } \\ 2\ 80\ \ \ \ \\ \underline{2\ 64}\ \ \ \ \\ 160\ \ \\ \underline{132}\ \ \\ 280\ \\ \underline{264}\ \\ 16\ \end{array} = 0.\overline{48}$$

13.
$$\begin{array}{r} 0.181 \\ 11\overline{)2.000} \\ \underline{1\ 1\ \ \ \ \ } \\ 90\ \ \ \\ \underline{88}\ \ \ \\ 20\ \\ \underline{11}\ \\ 9\ \end{array} \approx 0.18$$

15.
$$\begin{array}{r} 0.529 \\ 17\overline{)9.000} \\ \underline{8\ 5\ \ \ \ \ } \\ 50\ \ \ \\ \underline{34}\ \ \ \\ 160\ \\ \underline{153}\ \\ 7\ \end{array} \approx 0.53$$

17.
$$\begin{array}{r} 0.125 \\ 8\overline{)1.000} \\ \underline{8\ \ \ \ } \\ 20\ \ \\ \underline{16}\ \ \\ 40\ \\ \underline{40}\ \\ 0\ \end{array} \approx 0.13$$

19.
$$\begin{array}{r} 0.833 \\ 6\overline{)5.000} \\ \underline{4\ 8\ \ \ \ } \\ 20\ \ \\ \underline{18}\ \ \\ 20\ \\ \underline{18}\ \\ 2\ \end{array} \approx 0.83$$

21.
$$\begin{array}{r} 2.666 \\ 3\overline{)8.000} \\ \underline{6\ \ \ \ \ \ } \\ 2\ 0\ \ \ \\ \underline{1\ 8}\ \ \ \\ 20\ \ \\ \underline{18}\ \ \\ 20\ \\ \underline{18}\ \\ 2\ \end{array} \approx 2.67$$

23.
$$\begin{array}{r} 1.222 \\ 9\overline{)11.000} \\ \underline{9\ \ \ \ \ \ } \\ 2\ 0\ \ \ \\ \underline{1\ 8}\ \ \ \\ 20\ \ \\ \underline{18}\ \ \\ 20\ \\ \underline{18}\ \\ 2\ \end{array} \approx 1.22$$

Problem Set 4.7

25.
```
     0.916  ≈ 0.92
12)11.000
   10 8
      20
      12
       80
       72
        8
```

27. $3\frac{4}{15} \approx 3 + 0.27 = 3.27$
```
    0.266  ≈ 0.27
15)4.000
   3 0
   1 00
     90
    100
     90
     10
```

29. $6\frac{3}{8} = 6 + 0.375 = 6.375 \approx 6.38$
```
   0.375
8)3.000
  2 4
    60
    56
     40
     40
      0
```

31.
```
  0.8
5)4.0
  4 0
    0
```

33.
```
    0.4375  ≈ 0.44
16)7.0000
   6 4
     60
     48
     120
     112
       80
       80
        0
```

35.
```
      0.248  ≈ 0.25
125)31.000
    25 0
     6 00
     5 00
     1 000
     1 000
         0
```

37.
```
    1.48
25)37.00
   25
   12 0
   10 0
    2 00
    2 00
       0
```

39. $10\frac{7}{8} = 10 + 0.875 = 10.875$
```
   0.875  ≈ 10.88
8)7.000
  6 4
    60
    56
     40
     40
      0
```

41. $\frac{8}{9} \approx 0.9$; $\frac{8}{9} \approx 0.889$
```
  0.8888
9)8.0000
  7 2
    80
    72
     80
     72
      80
      72
       8
```

43. $\frac{11}{8} \approx 1.4$; $\frac{11}{8} = 1.375$
```
   1.375
8)11.000
   8
   3 0
   2 4
     60
     56
      40
      40
       0
```

45. $3\frac{7}{15} \approx 3.5$; $3\frac{7}{15} \approx 3.467$
```
   0.4666
15)7.0000
   6 0
   1 00
     90
    100
     90
     100
      90
      10
```

Problem Set 4.7

47. $\frac{21}{16} \approx 1.3$; $\frac{21}{16} \approx 1.313$

```
      1.3125
16)21.0000
   16
    5 0
    4 8
      20
      16
       40
       32
        80
        80
         0
```

49. $\frac{31}{125} \approx 0.2$; $\frac{31}{125} = 0.248$

```
       0.248
125)31.000
    25 0
     6 00
     5 00
     1 000
     1 000
         0
```

51. $\frac{62}{29} \approx 2.1$; $\frac{62}{29} \approx 2.138$

```
     2.1379
29)62.0000
   58
    4 0
    2 9
    1 10
      87
      230
      203
       270
       261
         9
```

53.

```
  0.2
5)1.0
  1 0
    0
```

55.

```
       0.517  ≈ 0.52
425)220.000
    212 5
      7 50
      4 25
      3 250
      2 975
        275
```

57.

```
     0.95
60)57.00
   54 0
    3 00
    3 00
       0
```

59. numerator, denominator

61. repeating

63. $7834 \div 35{,}619 \approx 0.220$

65. $\frac{7}{4} < \frac{9}{5}$, since $1.75 < 1.8$

```
  1.75       1.8
4)7.00     5)9.0
  4          5
  3 0        4 0
  2 8        4 0
    20         0
    20
     0
```

67. Seven and eight hundredths

69. Eleven and eight tenths

71. 84.07

73. $7\frac{69}{100}$

75. $8\frac{25}{1000} = 8\frac{1}{40}$

Chapter 4 Additional Exercises

1. Six and eight hundredths

3. Seven thousand, nine and six tenths

5. 761.05

7. $\frac{734}{1000} = \frac{367}{500}$

9. $0.06 > 0.009$

11. $7.999 < 8$

13. $5.1689 > 5.1679$

15. $0.8476 \approx 0.8$
$0.8476 \approx 0.85$
$0.8476 \approx 0.848$

Chapter 4 Additional Exercises

17. 22.8999 ≈ 22.9
22.8999 ≈ 22.90
22.8999 ≈ 22.900

19.
```
  402.86
   31.725
 +101.36
  535.945
```

21.
```
  402.86   ≈ 402.9
   31.725  ≈  31.7
 +101.36   ≈ 101.4
             536.0
```

23.
```
  460.48
 +275.34
  735.82
```
The new ticket price was $735.82.

25.
```
      15
    4 5̸  10
  7 5̸ 6̸ . 0̸
 -  4 9 . 3
  7 0 6 . 7
```

27.
```
       9
     7 1̸0 10
  0̸ . 8̸ 0̸ 0̸
 -0 . 6 1 4
  0 . 1 8 6
```

29.
```
  9.876  ≈  9.9
 -3.28   ≈ -3.3
            6.6
```

31.
```
      9   9  9 13
  6 1̸0 1̸0 1̸0 3̸   11
  7̸ 0̸ , 0̸ 0̸ 4̸ . 1̸
 -6 8 , 7 3 4 . 2
    1 , 2 6 9 . 9
```
Mark drove 1269.9 miles.

33.
```
    70.14
   × 8.02
   1 4028
  561 12
  562.5228
```

35.
```
     0.52
  ×0.0017
      364
      52
  0.000884
```

37. 18.95 × 4 ≈ 20 × 4 = 80
The cost is about $80.

39.
```
   92.78
 ×     4
  371.12
```
Sarah paid $371.12 for the tires.

41.
```
          3.414  ≈ 3.41
  0.28)0.95 600
          84
          11 6
          11 2
             40
             28
             120
             112
               8
```

43.
```
        3.617  ≈ 3.62
  9.4)34.0 000
       28 2
        5 8 0
        5 6 4
          1 60
            94
            660
            658
              2
```

45. 7.8 ÷ 100 = 0.078 ≈ 0.08

47.
```
           41.6
  8.70)361.92 0
       348 0
        13 92
         8 70
         5 22 0
         5 22 0
              0
```
Nazila worked 41.6 hours.

49.
$$\frac{125{,}000+245{,}000+432{,}500}{6} + \frac{175{,}500+324{,}000}{6} + \frac{143{,}000}{6} = 240{,}833.33$$
The average price of the houses was $240,833.33.

Chapter 4 Additional Exercises

51.

$\frac{11}{40} \approx 0.3$

$\frac{11}{40} \approx 0.28$

$\frac{11}{40} = 0.275$

$$\begin{array}{r} 0.275 \\ 40\overline{)11.000} \\ \underline{8\ 0} \\ 3\ 00 \\ \underline{2\ 80} \\ 200 \\ \underline{200} \\ 0 \end{array}$$

53.

$\frac{108}{85} \approx 1.3$

$\frac{108}{85} \approx 1.27$

$\frac{108}{85} \approx 1.271$

$$\begin{array}{r} 1.2705 \\ 85\overline{)108.0000} \\ \underline{85} \\ 23\ 0 \\ \underline{17\ 0} \\ 6\ 00 \\ \underline{5\ 95} \\ 500 \\ \underline{425} \\ 75 \end{array}$$

55.

$7\frac{2}{15} \approx 7.1$

$7\frac{2}{15} \approx 7.13$

$7\frac{2}{15} \approx 7.133$

$$\begin{array}{r} 0.1333 \\ 15\overline{)2.0000} \\ \underline{1\ 5} \\ 50 \\ \underline{45} \\ 50 \\ \underline{45} \\ 50 \\ \underline{45} \\ 5 \end{array}$$

57.

$5\frac{6}{13} \approx 5.5$

$5\frac{6}{13} \approx 5.46$

$5\frac{6}{13} \approx 5.462$

$$\begin{array}{r} 0.4615 \\ 13\overline{)6.0000} \\ \underline{5\ 2} \\ 80 \\ \underline{78} \\ 20 \\ \underline{13} \\ 70 \\ \underline{65} \\ 5 \end{array}$$

Chapter 4 Practice Test

1. Seven and thirty-four thousandths

3. $\frac{68}{100} = \frac{17}{25}$

5. $12.392 > 12.3916$

7. $7.1294 \approx 7.13$

9.

$$\begin{array}{r} 9.1 \\ 36.084 \\ +\ \underline{0.79} \\ 45.974 \end{array}$$

11.

$$\begin{array}{rcr} 7.84 & \approx & 8 \\ 0.76 & \approx & 1 \\ +\underline{15.265} & \approx & +\underline{15} \\ & & 24 \end{array}$$

13.

$$\begin{array}{r} \ \ 9 \\ 6\ \ 1\!\!/0\ 10 \\ 2\ \not{7}\ .\ \not{0}\ \not{0} \\ -\ \underline{6\ .\ 1\ 9} \\ 2\ 0\ .\ 8\ 1 \end{array}$$

15.

$$\begin{array}{r} 2.05 \\ \times\underline{9.78} \\ 1640 \\ 1\ 435 \\ \underline{18\ 45} \\ 20.0490 \end{array}$$

17. $3.4 \times 0.01 = 0.034$

19.

$$\begin{array}{r} 4.652 \\ 3.4\overline{)15.8\ 200} \\ \underline{13\ 6} \\ 2\ 2\ 2 \\ \underline{2\ 0\ 4} \\ 1\ 80 \\ \underline{1\ 70} \\ 100 \\ \underline{68} \\ 32 \end{array} \approx 4.65$$

21. $0.8 \div 100 = 0.008 \approx 0.01$

23. $6\frac{5}{12} \approx 6 + 0.417 \approx 6.417$

$$\begin{array}{r} 0.4166 \\ 12\overline{)5.0000} \\ \underline{4\ 8} \\ 20 \\ \underline{12} \\ 80 \\ \underline{72} \\ 80 \\ \underline{72} \\ 8 \end{array}$$

25.

$$\begin{array}{r} 472.85 \\ -\underline{194.29} \\ 278.56 \end{array}$$

Carlos paid $278.56 for clothing.

Chapter 4 Practice Test

27.

$$\begin{array}{r} 80. \\ 10.3\overline{)824.0} \\ \underline{824} \\ 0\ 0 \end{array}$$

Christine typed 80 words per minute.

Chapters 3 and 4 Cumulative Review

1.

$$\frac{2}{7}+\frac{3}{14}=\frac{2}{7}\cdot\frac{2}{2}+\frac{3}{14}=\frac{4}{14}+\frac{3}{14}=\frac{7}{14}=\frac{1}{2}$$

3.

$$\frac{11}{10}-\frac{4}{15}=\frac{11}{10}\cdot\frac{3}{3}-\frac{4}{15}\cdot\frac{2}{2}=\frac{33}{30}-\frac{8}{30}=\frac{25}{30}=\frac{5}{6}$$

5.

$$10\frac{3}{5}=\frac{(5\times 10)+3}{5}=\frac{53}{5}$$

7.

$$\frac{721}{8}=90\frac{1}{8}\qquad \begin{array}{r} 90 \\ 8\overline{)721} \\ \underline{72} \\ 01 \end{array}$$

9.

$$7\cdot\frac{8}{21}=\frac{7}{1}\cdot\frac{8}{21}=\frac{\cancel{7}}{1}\cdot\frac{8}{3\cdot\cancel{7}}=\frac{8}{3}=2\frac{2}{3}$$

11.

$$3\frac{2}{3}\cdot 6\frac{3}{5}=\frac{11}{3}\cdot\frac{33}{5}=\frac{11}{\cancel{3}}\cdot\frac{\cancel{3}\cdot 11}{5}=\frac{121}{5}=24\frac{1}{5}$$

13.

$$\frac{7}{4}\div\frac{5}{12}=\frac{7}{4}\cdot\frac{12}{5}=\frac{7}{\cancel{4}}\cdot\frac{3\cdot\cancel{4}}{5}=\frac{21}{5}=4\frac{1}{5}$$

15.

$$\begin{array}{rcl} 9\frac{3}{4}\cdot\frac{3}{3} & = & 9\frac{9}{12} \\ +8\frac{5}{6}\cdot\frac{2}{2} & = & +\ 8\frac{10}{12} \\ \hline & & 17\frac{19}{12}=18\frac{7}{12} \end{array}$$

17.

$$\begin{array}{rcl} 9\frac{5}{8}\cdot\frac{2}{2} & = & 9\frac{10}{16} \\ -3\frac{3}{16} & = & -3\frac{3}{16} \\ \hline & & 6\frac{7}{16} \end{array}$$

19.

$$\begin{array}{rcccl} 12\frac{7}{8}\cdot\frac{7}{7} & = & 12\frac{49}{56} & = & 11\frac{105}{56} \\ -\ 4\frac{13}{14}\cdot\frac{4}{4} & = & -\ 4\frac{52}{56} & = & -\ 4\frac{52}{56} \\ \hline & & & & 7\frac{53}{56} \end{array}$$

21.

$$\frac{17}{15}<\frac{5}{3}, \text{since } \frac{17}{15}<\frac{25}{15}$$

23.

$$\frac{7}{3}-\left(\frac{7}{4}-\frac{5}{8}\right)=\frac{7}{3}-\left(\frac{14}{8}-\frac{5}{8}\right)=\frac{7}{3}-\frac{9}{8}=\frac{56}{24}-\frac{27}{24}=\frac{29}{24}=1\frac{5}{24}$$

25.

$$34\frac{1}{2}\div 6\frac{3}{8}=\frac{69}{2}\div\frac{51}{8}=\frac{69}{2}\cdot\frac{8}{51}=\frac{\cancel{3}\cdot 23}{\cancel{2}}\cdot\frac{\cancel{2}\cdot 4}{\cancel{3}\cdot 17}=\frac{92}{17}=5\frac{7}{17}$$

Five pieces of ribbon can be cut with $\frac{7}{17}$ inches left over.

27.
$$\begin{array}{rcr} 24 & = & 23\frac{4}{4} \\ -15\frac{3}{4} & = & -15\frac{3}{4} \\ \hline & & 8\frac{1}{4} \end{array}$$

The difference in height was $8\frac{1}{4}$ pounds.

29. $9.38 = 9\frac{38}{100} = 9\frac{19}{50}$

31. $720.783 < 720.79$

33. $82.3095 \approx 82.31$

35.
$$\begin{array}{r} 16.7 \\ 2.03 \\ 4.8 \\ \hline 23.53 \end{array}$$

37.
$$\begin{array}{rcr} 0.69 & \approx & 1 \\ 5.75 & \approx & 6 \\ +3.20 & \approx & +3 \\ \hline & & \$10 \end{array}$$

39.
$$\begin{array}{r} 540.00 \\ -\ 23.19 \\ \hline 516.81 \end{array}$$

41. $5.6 \times 1000 = 5600$

43.
$$\begin{array}{r} 141.269 \\ 6.3\overline{)890.0000} \end{array}$$
$$\begin{array}{l} 63 \\ 260 \\ 252 \\ \quad 80 \\ \quad 63 \\ \quad 170 \\ \quad 126 \\ \quad\ 440 \\ \quad\ 378 \\ \quad\quad 620 \\ \quad\quad 567 \\ \quad\quad\ 53 \end{array}$$

45. $\dfrac{250.3+69.5}{2} = 159.9$

47.
$$\begin{array}{r} 3.8 \\ 6.4 \\ 1.2 \\ +\ 5.6 \\ \hline 17.0 \end{array}$$

John bought 17 pounds of ground beef.

49.
$$\begin{array}{r} 14.87 \\ +\ \ 8 \\ \hline 118.96 \end{array}$$

The total cost is \$118.96.

Problem Set 5.1

1. $\dfrac{7}{11}$

3. $\dfrac{90}{2} = \dfrac{45}{1}$

5. $\dfrac{12}{18} = \dfrac{2}{3}$

7. $\dfrac{319}{1000}$

9. $\dfrac{8}{15}$

11. $\dfrac{49}{98} = \dfrac{1}{2}$

13. $\dfrac{17}{51} = \dfrac{1}{3}$

15. $\dfrac{200}{100} = \dfrac{2}{1}$

17. $\dfrac{300}{199}$

19. $\dfrac{6\frac{3}{8}}{3} = \dfrac{\frac{51}{8}}{3} = \dfrac{51}{8} \div \dfrac{3}{1}$
$= \dfrac{51}{8} \bullet \dfrac{1}{3} = \dfrac{\cancel{3} \bullet 17}{8} \bullet \dfrac{1}{\cancel{3}}$
$= \dfrac{17}{8}$

21. $\dfrac{12}{5\frac{1}{3}} = \dfrac{12}{\frac{16}{3}} = \dfrac{12}{1} \div \dfrac{16}{3}$
$= \dfrac{12}{1} \bullet \dfrac{3}{16}$
$= \dfrac{3 \bullet \cancel{4}}{1} \bullet \dfrac{3}{\cancel{4} \bullet 4} = \dfrac{9}{4}$

Problem Set 5.1

23. $\dfrac{2\frac{5}{8}}{1\frac{7}{8}} = \dfrac{\frac{21}{8}}{\frac{15}{8}} = \dfrac{21}{8} \div \dfrac{15}{8}$
$= \dfrac{21}{8} \bullet \dfrac{8}{15}$
$= \dfrac{\cancel{3} \bullet 7}{\cancel{8}} \bullet \dfrac{\cancel{8}}{\cancel{3} \bullet 5} = \dfrac{7}{5}$

25. $\dfrac{2\frac{1}{2}}{3\frac{1}{2}} = \dfrac{\frac{5}{2}}{\frac{7}{2}} = \dfrac{5}{2} \div \dfrac{7}{2}$
$= \dfrac{5}{\cancel{2}} \bullet \dfrac{\cancel{2}}{7} = \dfrac{5}{7}$

27. $\dfrac{1\frac{2}{7}}{2\frac{1}{14}} = \dfrac{\frac{9}{7}}{\frac{29}{14}} = \dfrac{9}{7} \div \dfrac{29}{14}$
$= \dfrac{9}{7} \bullet \dfrac{14}{29} = \dfrac{9}{\cancel{7}} \bullet \dfrac{2 \bullet \cancel{7}}{29}$
$= \dfrac{18}{29}$

29. $\dfrac{6\frac{1}{5}}{8\frac{1}{10}} = \dfrac{\frac{31}{5}}{\frac{81}{10}} = \dfrac{31}{5} \div \dfrac{81}{10}$
$= \dfrac{31}{5} \bullet \dfrac{10}{81}$
$= \dfrac{31}{\cancel{5}} \bullet \dfrac{2 \bullet \cancel{5}}{81} = \dfrac{62}{81}$

31. $\dfrac{15}{25} = \dfrac{3}{5}$

33. $\dfrac{1200}{2200} = \dfrac{6}{11}$

35. $\dfrac{600}{425} = \dfrac{24}{17}$

37. $\dfrac{240}{240+330} = \dfrac{240}{570} = \dfrac{8}{19}$

39. $\dfrac{96-84}{84} = \dfrac{12}{84} = \dfrac{1}{7}$

41. comparison

43. $\dfrac{0.069}{0.123} = \dfrac{23}{41}$

45. $8.14 < 8.29$

47. $0.35 < 3.5$

49. $3.252 \approx 3.25$

51. $0.8643 \approx 0.86$

53. $9.0006 \approx 9.001$

55. $3.1598 \approx 3.160$

57. \$3.68 = 368 cents

59. \$0.36 = 36 cents

Problem Set 5.2

1. $\dfrac{\$0.89}{4 \text{ ounces}} = \dfrac{89 \text{ cents}}{4 \text{ ounces}}$
$= 22.25 \dfrac{\text{cents}}{\text{ounce}}$

3. $\dfrac{\$3.25}{28 \text{ ounces}} = \dfrac{325 \text{ cents}}{28 \text{ ounces}}$
$\approx 11.61 \dfrac{\text{cents}}{\text{ounce}}$

5. $\dfrac{\$2.69}{1 \text{ lb } 4 \text{ ounces}} = \dfrac{269 \text{ cents}}{20 \text{ ounces}}$
$= 13.45 \dfrac{\text{cents}}{\text{ounce}}$

7. $\dfrac{280 \text{ cm}}{8 \text{ seconds}} = 35 \dfrac{\text{cm}}{\text{second}}$

9. $\dfrac{8 \text{ quarts}}{16 \text{ pints}} = 0.5 \dfrac{\text{quarts}}{\text{pint}}$
$\dfrac{16 \text{ pints}}{8 \text{ quarts}} = 2 \dfrac{\text{pints}}{\text{quart}}$

11. $\dfrac{\$1.95}{5 \text{ pounds}} = \dfrac{195 \text{ cents}}{5 \text{ pounds}}$
$= 39 \dfrac{\text{cents}}{\text{pound}}$

13. $\dfrac{\$1153.84}{2 \text{ weeks}} = 576.92 \dfrac{\text{dollars}}{\text{week}}$

15. $\dfrac{372{,}568 \text{ miles}}{2 \text{ seconds}}$
$= 186{,}284 \dfrac{\text{miles}}{\text{second}}$

17. $\dfrac{252 \text{ words}}{4.5 \text{ minutes}} = 56 \dfrac{\text{words}}{\text{minute}}$

Problem Set 5.2

19. $\frac{\$2.40}{300 \text{ sheets}} = \frac{240 \text{ cents}}{300 \text{ sheets}}$
$= 0.8 \frac{\text{cents}}{\text{sheet}}$

21. $\frac{\$2.89}{3 \text{ rolls}} = \frac{289 \text{ cents}}{3 \text{ rolls}}$
$\approx 96.33 \frac{\text{cents}}{\text{roll}}$

23. $\frac{\$16.73}{3.5 \text{ yards}} = 4.78 \frac{\text{dollars}}{\text{yard}}$

25. Super Pen
$\frac{\$1.98}{6 \text{ pens}} = 0.33 \frac{\text{dollars}}{\text{pen}}$
Forever Pen
$\frac{\$1.75}{5 \text{ pens}} = 0.35 \frac{\text{dollars}}{\text{pen}}$
Super Pen has the lower unit price.

27. Orange Juice
$\frac{\$1.12}{16 \text{ ounces}} = 0.07 \frac{\text{dollars}}{\text{ounce}}$
Pineapple Juice
$\frac{\$3.15}{46 \text{ ounces}} \approx 0.068 \frac{\text{dollars}}{\text{ounce}}$
Pineapple juice has the lower unit price.

29. Kain Sugar
$\frac{\$2.38}{5 \text{ pounds}} = 0.476 \frac{\text{dollars}}{\text{pound}}$
SoSweet Sugar
$\frac{\$10.85}{25 \text{ pounds}} = 0.434 \frac{\text{dollars}}{\text{pound}}$
SoSweet sugar has the lower unit price.

31. Big O Stew
$\frac{\$0.79}{1 \text{ pound}} = 0.79 \frac{\text{dollars}}{\text{pound}}$
Bill Mill's Stew
$\frac{\$1.32}{1.5 \text{ pounds}} = 0.88 \frac{\text{dollars}}{\text{pound}}$
Big O Stew has the lower unit price.

33. Chocolate Cherries
$\frac{\$8.75}{4.3 \text{ pounds}} \approx 2.03 \frac{\text{dollars}}{\text{pound}}$
Peanut Brittle
$\frac{\$13.39}{6.7 \text{ pounds}} \approx 2.00 \frac{\text{dollars}}{\text{pound}}$
Peanut Brittle has the lower unit price.

35. rate

37. $\frac{\$2500 \text{ taxes}}{\$200{,}000 \text{ assessed value}}$
$= \frac{2500 \text{ dollars taxes}}{200 \text{ thousand dollars assessed value}}$
$= 12.5 \frac{\text{dollars taxes}}{\text{thousand dollars assessed values}}$

39. $\frac{\$112.50 - \$29.75}{25 \text{ plants}} = \frac{\$82.75}{25 \text{ plants}}$
$= 3.31 \frac{\text{dollars}}{\text{plant}}$

41.
$$\begin{array}{r} 307 \\ 6.2 \\ \underline{0.04} \\ 313.24 \end{array}$$

43.
$$\begin{array}{r} 3.6 \\ 0.08 \\ \underline{421.2} \\ 424.88 \end{array}$$

45. $0.89 + 3.56 + 2.4$
$\approx 1 + 4 + 2 \approx 7$

47. $16.2 + 5.93 + 101.7$
$\approx 16 + 6 + 102 = 124$

49. $3.40 + 0.89 + 4.74 + 1.59$
$\approx 3 + 1 + 5 + 2 = 11$
Mike's total bill is about \$11.

Problem Set 5.3

1. $\frac{8}{10} = \frac{80}{100}$

3. $\frac{22}{33} = \frac{24}{36}$

5. $\frac{75}{65} = \frac{15}{13}$

7. $\frac{16}{20} = \frac{64}{80}$

9. $\frac{2\frac{1}{5}}{9} = \frac{11}{45}$

11. $\frac{8}{16} = \frac{1}{2}$ Since $\frac{1}{2} = \frac{1}{2}$, the proportion is true.

Problem Set 5.3

13. $\frac{45}{25} = \frac{9}{5}$ Since $\frac{9}{5} = \frac{9}{5}$, the proportion is true.

15. Since $\frac{8}{9} \neq \frac{64}{81}$, the proportion is false.

17. $\frac{22}{18} = \frac{11}{9}$ $\frac{44}{36} = \frac{11}{9}$
Since $\frac{11}{9} = \frac{11}{9}$, the proportion is true.

19. $\frac{20}{14} = \frac{10}{7}$ $\frac{60}{51} = \frac{20}{17}$
Since $\frac{10}{7} \neq \frac{20}{17}$, the proportion is false.

21. $\frac{64}{72} = \frac{8}{9}$ $\frac{48}{54} = \frac{8}{9}$
Since $\frac{8}{9} = \frac{8}{9}$, the proportion is true.

23. Does 3 • 24 = 8 • 21?
Since 72 ≠ 168, the proportion is false.

25. Does 5 • 12 = 6 • 10?
Since 60 = 60, the proportion is true.

27. Does 8 • 42 = 6 • 56?
Since 336 = 336, the proportion is true.

29. Does 9 • 120 = 10 • 99?
Since 1080 ≠ 990, the proportion is false.

31. Does $\frac{2}{7} \cdot 3 = \frac{3}{14} \cdot 4$?
Since $\frac{6}{7} = \frac{6}{7}$, the proportion is true.

33. Does $\frac{3}{4} \cdot 9 = \frac{6}{5} \cdot 5$?
Since $\frac{27}{4} \neq \frac{6}{1}$, the proportion is false.

35. equal

37. cross

39. Does 52.9 × 57.2
= 14.3 × 210.6?
Since 3025.88 ≠ 3011.58, the proportion is false.

41. Does $4\frac{1}{4} \cdot 3 = 5\frac{5}{8} \cdot 2$?
$\frac{17}{4} \cdot \frac{3}{1} = \frac{45}{8} \cdot \frac{2}{1}$?
Since $\frac{51}{4} \neq \frac{45}{4}$, the proportion is false.

43.
$$9 \cdot 3\frac{1}{3} = \frac{9}{1} \cdot \frac{10}{3} = \frac{\cancel{3} \cdot 3}{1} \cdot \frac{10}{\cancel{3}} = 30$$

45.
$$9\frac{1}{2} \div 5\frac{3}{7} = \frac{19}{2} \div \frac{38}{7} = \frac{19}{2} \cdot \frac{7}{38} = \frac{\cancel{19}}{2} \cdot \frac{7}{2 \cdot \cancel{19}} = \frac{7}{4} = 1\frac{3}{4}$$

47.
$$8\frac{1}{6} \div 7 = \frac{49}{6} \div \frac{7}{1} = \frac{49}{6} \cdot \frac{1}{7} = \frac{\cancel{7} \cdot 7}{6} \cdot \frac{1}{\cancel{7}} = \frac{7}{6} = 1\frac{1}{6}$$

49.
$$8 \cdot 11\frac{3}{4} = \frac{8}{1} \cdot \frac{47}{4} = \frac{2 \cdot \cancel{4}}{1} \cdot \frac{47}{\cancel{4}} = 94$$
Juan earns $94 in $11\frac{3}{4}$ hrs.

51.
$$6 \cdot 2\frac{1}{4} = \frac{6}{1} \cdot \frac{9}{4} = \frac{\cancel{2} \cdot 3}{1} \cdot \frac{9}{\cancel{2} \cdot 2} = \frac{27}{2} = 13\frac{1}{2}$$
There are $13\frac{1}{2}$ servings.

Problem Set 5.4

1. $\frac{x}{5} = \frac{6}{10}$ $10 \cdot x = 5 \cdot 6$
$x = \frac{30}{10}$ $x = 3$

Problem Set 5.4

3. $\frac{x}{7} = \frac{16}{14}$ $\quad 14 \cdot x = 7 \cdot 16$

$x = \frac{112}{14}$ $\quad x = 8$

5. $\frac{12}{t} = \frac{6}{5}$ $\quad 12 \cdot 5 = 6 \cdot t$

$\frac{60}{6} = t$ $\quad 10 = t$

7. $\frac{8}{24} = \frac{x}{9}$ $\quad 8 \cdot 9 = 24 \cdot x$

$\frac{72}{24} = x$ $\quad 3 = x$

9. $\frac{4}{5} = \frac{8}{n}$ $\quad 4 \cdot n = 5 \cdot 8$

$n = \frac{40}{4}$ $\quad n = 10$

11. $\frac{7}{11} = \frac{3}{x}$ $\quad 7 \cdot x = 3 \cdot 11$

$x = \frac{33}{7}$ $\quad x = 4\frac{5}{7}$

13. $\frac{11}{6} = \frac{x}{12}$ $\quad 11 \cdot 12 = 6 \cdot x$

$\frac{132}{6} = x$ $\quad 22 = x$

15. $\frac{2}{5} = \frac{5}{y}$ $\quad 2 \cdot y = 5 \cdot 5$

$y = \frac{25}{2}$ $\quad y = 12\frac{1}{2}$

17. $\frac{15}{9} = \frac{x}{7}$ $\quad 15 \cdot 7 = 9 \cdot x$

$\frac{105}{9} = x$ $\quad 11\frac{2}{3} = x$

19. $\frac{13}{y} = \frac{7}{2}$ $\quad 13 \cdot 2 = 7 \cdot y$

$\frac{26}{7} = y$ $\quad 3\frac{5}{7} = y$

21. $\frac{0.3}{0.5} = \frac{n}{25}$ $\quad 0.3 \times 25 = 0.5 \times n$

$\frac{7.5}{0.5} = n$ $\quad 15 = n$

23. $\frac{75}{100} = \frac{x}{40}$ $\quad 75 \cdot 40 = 100 \cdot x$

$\frac{3000}{100} = x$ $\quad 30 = x$

25. $\frac{x}{24} = \frac{\frac{1}{4}}{5}$ $\quad 5 \cdot x = 24 \cdot \frac{1}{4}$

$x = \frac{6}{5}$ $\quad x = 1\frac{1}{5}$

27. $\frac{n}{10} = \frac{\frac{3}{5}}{\frac{3}{8}}$ $\quad \frac{3}{8} \cdot n = 10 \cdot \frac{3}{5}$

$n = \frac{\frac{30}{5}}{\frac{3}{8}}$ $\quad n = \frac{30}{5} \cdot \frac{8}{3}$

$n = \frac{240}{15}$ $\quad n = 16$

29. $\frac{1}{100} = \frac{x}{25}$ $\quad 1 \cdot 25 = 100 \cdot x$

$\frac{25}{100} = x$ $\quad \frac{1}{4} = x$

31. $\frac{0.3}{x} = \frac{3.2}{0.8}$

$0.3 \times 0.8 = 3.2 \times x$

$\frac{0.24}{3.2} = x$ $\quad 0.075 = x$

33. $\frac{240}{6} = \frac{x}{10}$ $\quad 240 \cdot 10 = 6 \cdot x$

$\frac{2400}{6} = x$ $\quad 400 = x$

Kevin earns $400 for working 10 days.

35. $\frac{6}{16} = \frac{x}{48}$ $\quad 6 \cdot 48 = 16 \cdot x$

$\frac{288}{16} = x$ $\quad 18 = x$

Aaron gets 18 hits in 48 games.

37. $\frac{6}{75} = \frac{8}{x}$ $\quad 6 \cdot x = 75 \cdot 8$

$x = \frac{600}{6}$ $\quad x = 100$

Eight pencils cost 100 cents or $1.00.

39. $\frac{6}{4200} = \frac{x}{3500}$

$6 \cdot 3500 = 4200 \cdot x$

$\frac{21{,}000}{4200} = x$ $\quad 5 = x$

Five pounds of grass seed are needed.

Problem Set 5.4

41. $\dfrac{\frac{3}{4}}{3} = \dfrac{x}{7}$ $\qquad \dfrac{3}{4} \cdot 7 = 3 \cdot x$

$\dfrac{\frac{21}{4}}{3} = x$ $\qquad \dfrac{21}{4} \div \dfrac{3}{1} = x$

$\dfrac{\not{3} \cdot 7}{4} \cdot \dfrac{1}{\not{3}} = x$

$\dfrac{7}{4} = x$ $\qquad 1\dfrac{3}{4} = x$

The material needed is $1\frac{3}{4}$ yards.

43. $\dfrac{3}{2.37} = \dfrac{n}{7.11}$

$3 \times 7.11 = 2.37 \times n$

$\dfrac{21.33}{2.37} = n \qquad 9 = n$

Nine cans can be bought.

45. $\dfrac{7}{280} = \dfrac{16}{x} \qquad 7 \cdot x = 280 \cdot 16$

$x = \dfrac{4480}{7} \qquad x = 640$

Jake will travel 640 miles.

47. $\dfrac{39}{31} = \dfrac{n}{145.7}$

$39 \times 145.7 = 31 \times n$

$\dfrac{5682.3}{31} = n \quad 183.3 = n$

The grams of cereal would be 183.3.

49. $\dfrac{75}{5} = \dfrac{180}{x} \qquad 75 \cdot x = 5 \cdot 180$

$x = \dfrac{900}{75} \qquad x = 12$

There would be 12 defective.

51. $\dfrac{3.5}{100} = \dfrac{n}{40{,}000}$

$3.5 \times 40{,}000 = 100 \times n$

$\dfrac{140{,}000}{100} = n; \; 1400 = n$

There would be 1400 pounds of salt.

53. $\dfrac{28{,}000}{4200} = \dfrac{30{,}000}{x}$

$28{,}000 \cdot x = 4200 \cdot 30{,}000$

$x = \dfrac{126{,}000{,}000}{28{,}000}; \; x = 4500$

Kendrick University will have 4500 students living in dormitories.

55. like, like

57. $\dfrac{285}{630} = \dfrac{114}{x}$

$285 \cdot x = 630 \cdot 114$

$x = \dfrac{71{,}820}{285} \qquad x = 252$

59. $\dfrac{902}{617} = \dfrac{p}{181}$

$902 \cdot 181 = 617 \cdot p$

$\dfrac{163{,}262}{617} = p$

$264.61 \approx p$

61. $\dfrac{528}{32} = 16.5$

The amount of gasoline used on the trip was 16.5 gallons.

$16.5 \times 1.085 = 17.9025$

The total cost of the gasoline was about $17.90.

63.
```
     9  17 11
   3 10  7 1 10
  6 4 0 . 8 2 0
 -3 2 7 . 9 5 4
  3 1 2 . 8 6 6
```

65.
```
  3 17
  4 7 . 3 5
 -2 9 . 0 0
  1 8 . 3 5
```

67.
```
     9  9   9  9  9
  4 10 10  10 10 1010
  5 0 0 . 0 0 0 0
 -  8 5 . 9 9 9 9
  4 1 4 . 0 0 0 1
```

69.
```
     9  9
  8  10 10 12
  9 . 0 0 2
 -0 . 7 5 9
  8 . 2 4 3
```

71.
```
   76.30          8 12
    9.25        4 9 2 . 9 4
  324.00       -4 2 7 . 5 0
 + 83.39          6 5 . 4 4
  492.94
```
John needs $65.44.

Chapter 5 Additional Exercises

1. $\dfrac{150}{3} = \dfrac{50}{1}$

Chapter 5 Additional Exercises

3. $\frac{11}{9}$

5. $\frac{32}{36} = \frac{8}{9}$

7. $\frac{200}{126} = \frac{100}{63}$

9. $\frac{8\frac{1}{8}}{5} = \frac{\frac{65}{8}}{5} = \frac{65}{8} \div \frac{5}{1}$
$= \frac{\not{5} \bullet 13}{8} \bullet \frac{1}{\not{5}} = \frac{13}{8}$

11. $\frac{2200}{240{,}000} = \frac{11}{1200}$

13. $\frac{64 \text{ ounces}}{4 \text{ pounds}} = 16 \frac{\text{ounces}}{\text{pound}}$

15. $\frac{950 \text{ miles}}{3.75 \text{ hours}} \approx 253.3 \frac{\text{miles}}{\text{hour}}$

17. Nut'n Wheat
$\frac{\$0.98}{24 \text{ ounces}} \approx 0.04 \frac{\text{dollars}}{\text{ounce}}$
Ohio
$\frac{\$0.83}{16 \text{ ounces}} \approx 0.05 \frac{\text{dollars}}{\text{ounce}}$
The Nut'n Wheat bread has the lower unit price.

19. Oregon
$\frac{\$0.69}{5 \text{ pounds}} \approx 0.138 \frac{\text{dollars}}{\text{pounds}}$
Maine
$\frac{\$1.79}{10 \text{ pounds}} \approx 0.179 \frac{\text{dollars}}{\text{pounds}}$
The Oregon potatoes have the lower unit price.

21. $\frac{5}{9} = \frac{25}{45}$

23. $\frac{63}{72} = \frac{7}{8}$ $\quad \frac{42}{49} = \frac{6}{7}$
Since $\frac{7}{8} \neq \frac{6}{7}$, the proportion is false.

25. Does $12 \bullet 65 = 15 \bullet 52$?
Since $780 = 780$, the proportion is true.

27. Does $5\frac{1}{4} \bullet 7 = 6\frac{1}{8} \bullet 6$?
$\frac{21}{4} \bullet \frac{7}{1} = \frac{49}{8} \bullet \frac{6}{1}$?
Since $\frac{147}{4} = \frac{147}{4}$, the proportion is true.

29. $\frac{7}{x} = \frac{8}{20}$ $\quad 7 \bullet 20 = 8 \bullet x$
$\frac{140}{8} = x$ $\quad 17\frac{1}{2} = x$

31. $\frac{x}{7.2} = \frac{3.5}{10.5}$
$10.5 \times x = 7.2 \times 3.5$
$x = \frac{25.2}{10.5}$ $\quad x = 2.4$

33. $\frac{4}{425} = \frac{9}{x}$ $\quad 4 \bullet x = 425 \bullet 9$
$x = \frac{3825}{4}$ $\quad x = 956.25$
The cost is \$956.25.

35. $\frac{26.7}{1} = \frac{x}{725}$
$26.7 \times 725 = 1 \bullet x$
$19{,}357.5 = x$
Kathy gets 19,357.5 cents or about \$193.58.

37. $\frac{1800}{1800-36} = \frac{700}{x}$
$\frac{1800}{1764} = \frac{700}{x}$
$1800 \bullet x = 1764 \bullet 700$
$x = \frac{1{,}234{,}800}{1800}$
$x = 686$
The expected number of not irregular towels is 686.

Chapter 5 Practice Test

1. $\frac{24}{36} = \frac{2}{3}$

3. $\frac{55 \text{ miles}}{65 \text{ minutes}} = \frac{11}{13} \frac{\text{miles}}{\text{minute}}$

5. $\frac{\$9.12}{48 \text{ ounces}} = \frac{912 \text{ cents}}{48 \text{ ounces}}$
$= 19 \frac{\text{cents}}{\text{ounce}}$

Chapter 5 Practice Test

7. Does $21 \bullet 83 = 49 \bullet 36$?
Since $1743 \neq 1764$, the proportion is not true.

9. $\frac{8}{11} = \frac{5}{x}$ $\quad 8 \bullet x = 5 \bullet 11$
$x = \frac{55}{8}$ $\quad x = 6\frac{7}{8}$

11. $\frac{2}{1.29} = \frac{n}{15.48}$
$2 \times 15.48 = 1.29 \times n$
$\frac{30.96}{1.29} = n \quad 24 = n$
The number of bottles of juice that can be purchased is 24.

Problem Set 6.1

1. $44\% = \frac{44}{100}$
$44\% = 44 \times 0.01$
$44\% = 44 \times \frac{1}{100}$

3. $8.2\% = \frac{8.2}{100}$
$8.2\% = 8.2 \times 0.01$
$8.2\% = 8.2 \times \frac{1}{100}$

5. $91\% = \frac{91}{100}$
$91\% = 91 \times 0.01$
$91\% = 91 \times \frac{1}{100}$

7. $130\% = \frac{130}{100}$
$130\% = 130 \times 0.01$
$130\% = 130 \times \frac{1}{100}$

9. $52\% = 0.52$

11. $38.5\% = 0.385$

13. $7.5\% = 0.075$

15. $68.34\% = 0.6834$

17. $72\% = 0.72$

19. $0.3\% = 0.003$

21. $0.98\% = 0.0098$

23. $140\% = 1.40$

25. $25\% = 0.25$

27. $4\% = 0.04$

29. $0.42 = 42\%$

31. $6.79 = 679\%$

33. $0.285 = 28.5\%$

35. $6.00 = 600\%$

37. $1 = 100\%$

39. $0.034 = 3.4\%$

41. $8.23 = 823\%$

43. $0.007 = 0.7\%$

45. $0.3 = 30\%$

47. $0.06 = 6\%$

49. $0.07190 = 7.190\%$

51. hundred

53. two, right

55. $\frac{7}{50} = \frac{14}{100}$
14% of the students were absent

57. $\frac{2}{3} \approx 0.7$
$\frac{2}{3} \approx 0.67$
$\frac{2}{3} \approx 0.667$

```
   0.6666
3)2.0000
  1 8
    20
    18
     20
     18
      20
      18
       2
```

59. $\frac{9}{13} \approx 0.7$
$\frac{9}{13} \approx 0.69$
$\frac{9}{13} \approx 0.692$

```
    0.6923
13)9.0000
   7 8
   1 20
   1 17
      30
      26
       40
       39
        1
```

Problem Set 6.1

61. $7\frac{5}{16} \approx 7.3$

$7\frac{5}{16} \approx 7.31$

$7\frac{5}{16} \approx 7.313$

```
    0.3125
16)5.0000
   4 8
     20
     16
      40
      32
       80
       80
        0
```

63. $4\frac{5}{11} \approx 4.5$

$4\frac{5}{11} \approx 4.45$

$4\frac{5}{11} \approx 4.455$

```
    0.4545
11)5.0000
   4 4
     60
     55
      50
      44
       60
       55
        5
```

Problem Set 6.2

1. $\frac{1}{8} = 0.125$

$= 12.5\%$

```
   0.125
8)1.000
    8
    20
    16
     40
     40
      0
```

3. $\frac{3}{7} \approx 0.429$

$\approx 42.9\%$

```
   0.4285
7)3.0000
  2 8
    20
    14
     60
     56
      40
      36
       4
```

5. $\frac{4}{9} \approx 0.444$

$\approx 44.4\%$

```
   0.4444
9)4.0000
  3 6
    40
    36
     40
     36
      40
      36
       4
```

7. $\frac{73}{100} = 0.73 = 73\%$

9. $\frac{4}{7} \approx 0.571$

$\approx 57.1\%$

```
   0.5714
7)4.0000
  3 5
    50
    49
     10
      7
     30
     28
      2
```

11. $\frac{3}{5} = 0.60 = 60\%$

13. $\frac{4}{25} = 0.16 = 16\%$

$\frac{4}{25} \bullet \frac{4}{4} = \frac{16}{100} = 0.16$

15. $\frac{6}{7} \approx 0.857$

$\approx 85.7\%$

```
   0.8571
7)6.0000
  5 6
    40
    35
     50
     49
      10
       7
       3
```

17. $\frac{89}{100} = 0.89 = 89\%$

19. $\frac{1}{4} = 0.25 = 25\%$

$\frac{1}{4} \bullet \frac{25}{25} = \frac{25}{100} = 0.25$

21. $7\frac{5}{8} = 7.625$

$= 762.5\%$

```
   0.625
8)5.000
  4 8
    20
    16
     40
     40
      0
```

23. $\frac{5}{2} = 2.5 = 250\%$

```
   2.5
2)5.0
  4
  1 0
  1 0
    0
```

Problem Set 6.2

25. $\frac{17}{25} = 0.68 = 68\%$

$$\begin{array}{r} 0.68 \\ 25\overline{)17.00} \\ \underline{15\ 0} \\ 2\ 00 \\ \underline{2\ 00} \\ 0 \end{array}$$

27. $51\% = \frac{51}{100}$

29. $42\% = \frac{42}{100} = \frac{21}{50}$

31. $140\% = \frac{140}{100} = \frac{7}{5}$

33. $125\% = \frac{125}{100} = \frac{5}{4}$

35. $87.5\% = \frac{87.5}{100} = \frac{7}{8}$

37. $2.75\% = \frac{2.75}{100} = \frac{11}{400}$

39. $8\frac{1}{3}\% = \frac{25}{3} \times \frac{1}{100} = \frac{1}{12}$

41. $83\frac{1}{3}\% = \frac{250}{3} \times \frac{1}{100} = \frac{5}{6}$

43. $45\% = \frac{45}{100} = \frac{9}{20}$

45. $\frac{79}{400} = 0.1975 = 19.75\% \approx 19.8\%$

47. $\frac{429}{315} \approx 1.362 = 136.2\%$

49. $346\% = \frac{346}{100} = \frac{173}{50}$

51. $\frac{370}{1000} = \frac{37}{100} = 37\%$

53. $\frac{3}{5} = \frac{60}{100}$

55. $\frac{3}{2\frac{1}{4}} = \frac{4}{3}$

57. $\frac{7}{25} = \frac{a}{100}$ $\quad 7 \bullet 100 = 25 \bullet a$

$\frac{700}{25} = a$ $\quad 28 = a$

59. $\frac{90}{120} = \frac{p}{100}$ $\quad 90 \bullet 100 = 120 \bullet p$

$\frac{9000}{120} = p$ $\quad 75 = p$

61. $\frac{9}{b} = \frac{54}{100}$ $\quad 9 \bullet 100 = 54 \bullet b$

$\frac{900}{54} = b$ $\quad 16\frac{2}{3} = b$

Problem Set 6.3

1. $\frac{a}{25} = \frac{16}{100}$ $\quad 100 \bullet a = 25 \bullet 16$

$a = \frac{400}{100}$ $\quad a = 4$

3. $\frac{7}{b} = \frac{28}{100}$ $\quad 7 \bullet 100 = 28 \bullet b$

$\frac{700}{28} = b$ $\quad 25 = b$

5. $\frac{20}{24} = \frac{p}{100}$ $\quad 20 \bullet 100 = 24 \bullet p$

$\frac{2000}{24} = p$ $\quad 83.3 \approx p$

7. $\frac{25}{b} = \frac{5}{100}$ $\quad 25 \bullet 100 = 5 \bullet b$

$\frac{2500}{5} = b$ $\quad 500 = b$

9. $\frac{a}{65} = \frac{37}{100}$

11. $\frac{6.7}{8.2} = \frac{p}{100}$

13. $\frac{7}{b} = \frac{15}{100}$

15. $\frac{a}{148} = \frac{25}{100}$ $\quad 100 \bullet a = 25 \bullet 148$

$a = \frac{3700}{100}$ $\quad a = 37$

17. $\frac{a}{30} = \frac{42}{100}$ $\quad 100 \bullet a = 30 \bullet 42$

$a = \frac{1260}{100}$ $\quad a = 12.6$

19. $\frac{a}{16} = \frac{73}{100}$ $\quad 100 \bullet a = 16 \bullet 73$

$a = \frac{1168}{100}$ $\quad a = 11.68 \approx 11.7$

Problem Set 6.3

21. $\frac{6}{b} = \frac{15}{100}$ $\quad 6 \cdot 100 = 15 \cdot b$

$\frac{600}{15} = b$ $\quad 40 = b$

23. $\frac{17}{b} = \frac{30}{100}$ $\quad 17 \cdot 100 = 30 \cdot b$

$\frac{1700}{30} = b$ $\quad 56.7 \approx b$

25. $\frac{42}{b} = \frac{25}{100}$ $\quad 42 \cdot 100 = 25 \cdot b$

$\frac{4200}{25} = b$ $\quad 168 = b$

27. $\frac{14}{56} = \frac{p}{100}$ $\quad 14 \cdot 100 = 56 \cdot p$

$\frac{1400}{56} = p$ $\quad 25 = p$

29. $\frac{3.6}{18} = \frac{p}{100}$ $\quad 3.6 \times 100 = 18 \times p$

$\frac{360}{18} = p$ $\quad 20 = p$

31. $\frac{105}{100} = \frac{p}{100}$

$105 \cdot 100 = 100 \cdot p$

$\frac{10500}{100} = p$ $\quad 105 = p$

33. $\frac{a}{85} = \frac{30}{100}$ $\quad 100 \cdot a = 85 \cdot 30$

$a = \frac{2550}{100}$ $\quad a = 25.5$

35. $\frac{22}{11} = \frac{p}{100}$ $\quad 22 \cdot 100 = 11 \cdot p$

$\frac{2200}{11} = p$ $\quad 200 = p$

37. $\frac{4}{b} = \frac{16.5}{100}$ $\quad 4 \cdot 100 = 16.5 \cdot b$

$\frac{400}{16.5} = 6$ $\quad 24.2 \approx b$

39. $\frac{a}{14,300} = \frac{57}{100}$

$100 \cdot a = 57 \cdot 14,300$

$a = \frac{815,100}{100}$ $\quad a = 8151$

The number of people who voted for the school bonds was 8151.

41. $\frac{8400}{b} = \frac{32}{100}$

$8400 \cdot 100 = 32 \cdot b$

$\frac{840,000}{32} = b$

$26,250 = b$

Noah's salary is \$26,250.

43. $\frac{6}{24} = \frac{p}{100}$ $\quad 6 \cdot 100 = 24 \cdot p$

$\frac{600}{24} = P$ $\quad 25 = P$

Cynthia spends 25% of the day studying.

45. $\frac{a}{38,000} = \frac{14}{100}$

$100 \cdot a = 38,000 \cdot 14$

$a = \frac{532,000}{100}$

$a = 5320$

Lecia plans to spend \$5320 for rent.

47. $\frac{3}{75} = \frac{p}{100}$ $\quad 3 \cdot 100 = 75 \cdot p$

$\frac{300}{75} = p$ $\quad 4 = p$

Nicholas got 4% of the questions wrong.

49. $\frac{22,500}{b} = \frac{150}{100}$

$22,500 \cdot 100 = 150 \cdot b$

$\frac{2,250,000}{150} = b$

$15,000 = b$

The population was 15,000.

51. percent, %

53. part

55. $\frac{16.50}{89.74} = \frac{p}{100}$

$16.50 \cdot 100 = 89.74 \cdot p$

$\frac{1650}{89.74} = p$ $\quad 18.39 \approx p$

18.39%

57. $\frac{a}{22,000} = \frac{11.3}{100}$

$100 \cdot a = 22,000 \cdot 11.3$

$a = \frac{248,600}{100}$

$a = 2486$ $\quad$ \$2486

Problem Set 6.3

59. $\dfrac{3\frac{1}{5}}{b} = \dfrac{9\frac{1}{10}}{100}$

$\dfrac{16}{5} \bullet \dfrac{100}{1} = \dfrac{91}{10} \bullet b$

$\dfrac{1600}{5} \bullet \dfrac{10}{91} = b$

$35\dfrac{15}{91} = b$

61. $\dfrac{64}{70} > \dfrac{27}{30}$; since $\dfrac{192}{210} > \dfrac{189}{210}$

Ed got the greater percentage of questions correct.

63.
```
            90.243   ≈ 9.24
0.82)74.00 000
  --'   --'
    73 8
       20 0
       16 4
          360
          246
           74
```

65. $76.1 \div 0.01 = 7610$

67.
```
           6.785
1.12)7.60 000
  --'  --'
    6 72
      88 0
      78 4
       9 60
       8 96
         640
         560
          80
```

69. $\dfrac{17.4 + 18.09 + 8 + 10.1}{4}$

$= 13.3975 \approx 13.40$

71. $\dfrac{90 + 35.8 + 46.24}{3} \approx 57.35$

Problem Set 6.4

1. $a = 59\% \bullet 11$

3. $18 = 32\% \bullet b$

5. $39 = p\% \bullet 42$

7. $a = 15\% \bullet 40$
$a = 0.15 \bullet 40$
$a = 6$

9. $a = 120\% \bullet 50$
$a = 1.20 \bullet 50$
$a = 60$

11. $a = 40\% \bullet 72$
$a = 0.40 \bullet 72$
$a = 28.8$

13. $36 = 3\% \bullet b$
$36 = 0.03 \bullet b$
$\dfrac{36}{0.03} = b$
$1200 = b$

15. $35 = 50\% \bullet b$
$35 = 0.50 \bullet b$
$\dfrac{35}{0.50} = b$
$70 = b$

17. $52 = 100\% \bullet b$
$52 = 1.00 \bullet b$
$\dfrac{52}{1.00} = b$
$52 = b$

19. $9 = p\% \bullet 60$
$9 = p \bullet 0.01 \bullet 60$
$9 = p \bullet 0.6$
$\dfrac{9}{0.6} = p$
$15 = p$

21. $120 = p\% \bullet 80$
$120 = p \bullet 0.01 \bullet 80$
$120 = p \bullet 0.8$
$\dfrac{120}{0.8} = p$
$150 = p$

23. $8 = p\% \bullet 34$
$8 = p \bullet 0.01 \bullet 34$
$8 = p \bullet 0.34$
$\dfrac{8}{0.34} = p$
$23.5 \approx p$

25. $a = 28\% \bullet 50$
$a = 0.28 \bullet 50$
$a = 14$

27. $5.2 = p\% \bullet 6.5$
$5.2 = p \bullet 0.01 \bullet 6.5$
$5.2 = P \bullet 0.065$
$\dfrac{5.2}{0.065} = p$
$80 = p$

29. $a = 28\% \bullet 99$
$a = 0.28 \bullet 99$
$a = 27.72$
$a \approx 27.7$

Problem Set 6.4

31. $6720 = p\% \cdot 48{,}000$
$6720 = p \cdot 0.01 \cdot 48{,}000$
$6720 = p \cdot 480$
$\frac{6720}{480} = p$
$14 = p$
14% of the donations was spent for administrative expenses.

33. $a = 8\% \cdot 250$
$a = 0.08 \cdot 250$
$a = 20$
They received 20 false alarms.

35. $228 = 120\% \cdot b$
$228 = 1.20 \cdot b$
$\frac{228}{1.20} = b$
$190 = b$
Last year's snowfall was 190 inches.

37. $a = 11\% \cdot (650-400)$
$a = 0.11 \cdot 250$
$a = 27.5$
Leah should pay $27.50.

39. $(5200-3380) = p\% \cdot 5200$
$1820 = p \cdot 0.01 \cdot 5200$
$1820 = p \cdot 52$
$\frac{1820}{52} = p$
$35 = p$
35% of the students are vocational students.

41. $2052 = 90\% \cdot b$
$2052 = 0.90 \cdot b$
$\frac{2052}{0.90} = b$
$2280 = b$
$2280-2052 = 228$
The difference in attendance between the two years is 228.

43. base

45. $a = 9.48\% \cdot 732.09$
$a = 0.0948 \cdot 732.09$
$a \approx 69.40$

47. $216 = 46\% \cdot b$
$216 = 0.46 \cdot b$
$\frac{216}{0.46} = b$
$469.57 \approx b$

49. $20 = 6\frac{1}{4}\% \cdot b$
$20 = 0.0625 \cdot b$
$\frac{20}{0.0625} = b$
$320 = b$

51.
```
   86.2
  × 4.6
  51 72
 344 8
 396.52
```

53.
```
    22.3
   ×5.92
     446
  20 07
 111 5
 132.016
```

55. $5.3 \times 0.01 = 0.053$

57. $8 \times 94.32 \approx 8 \times 90 = 720$
The answer is about $720.

59. $10 \times 86.85 \approx 10 \times 90 = 900$
The answer is about $900.

61.
```
  21.82
 ×    6
 130.92
```
Carlos paid $130.92 for the shirts.

Problem Set 6.5

1. Sales tax = 4% • $100
= 0.04 • $100 = $4
Total price = $100 + $4
= $104

3. Sales tax = 8% • $70
= 0.08 • $70 = $5.60
Total price = $70 + $5.60
= $75.60

5. Sales tax = 5% • $245
= 0.05 • $245 = $12.25
Total price = $245 + $12.25
= $257.25

7. Sales tax = 7% • $12
= 0.07 • $12 = $0.84
Total price = $12 + $0.84
= $12.84

Problem Set 6.5

9. Sales tax = 6.5% • \$74
   = 0.065 • \$74 = \$4.81
   Total price = \$74 + \$4.81
   = \$78.81

11. Commission = 10% • \$200
    = 0.10 • \$200 = \$20

13. Commission = 28% • \$1000
    = 0.28 • \$1000 = \$280

15. Commission = 4% • \$6725
    = 0.04 • \$6725 = \$269

17. Commission = 9% • \$30,000
    = 0.09 • \$30,000
    = \$2700

19. Commission = 6% • \$134,000
    = 0.06 • \$134,000
    = \$8040

21. Discount = 20% • \$100
    = 0.20 • \$100 = \$20
    Sale price = \$100 - \$20
    = \$80

23. Discount = 5% • \$640
    = 0.05 • \$640 = \$32
    Sale price = \$640 - \$32
    = \$608

25. Discount = 40% • \$24.30
    = 0.40 • \$24.30
    = \$9.72
    Sale price = \$24.30 - \$9.72
    = \$14.58

27. Discount = 10% • \$16.85
    = 0.10 • \$16.85
    ≈ \$1.69
    Sale price = \$16.85 - \$1.69
    = \$15.16

29. Discount = 30% • \$8000
    = 0.30 • \$8000 = \$2400
    Sale price = \$8000 - \$2400
    = \$5600

31. Amount of increase
    = \$100 - \$80 = \$20
    Percent increase
    $= \frac{20 \cdot 100}{80} = 25$

33. Amount of increase
    = \$218 - \$200 = \$18
    Percent increase
    $= \frac{18 \cdot 100}{200} = 9$

35. Amount of increase
    = \$125 - \$100 = \$25
    Percent increase
    $= \frac{25 \cdot 100}{100} = 25$

37. Amount of increase
    = \$500 - \$400 = \$100
    Percent increase
    $= \frac{100 \cdot 100}{500} = 20$

39. Amount of increase
    = \$40 - \$32 = \$8
    Percent increase
    $= \frac{8 \cdot 100}{40} = 20$

41. Amount of increase
    = \$340 - \$320 = \$20
    Percent increase
    $= \frac{20 \cdot 100}{340} \approx 5.9$

43. Sales tax = 7% • \$14,000
    = 0.07 • \$14,000
    = \$980
    Total price = \$14,000 + \$980
    = \$14,980

45. Sales tax rate
    $= \frac{\$19.20}{\$240} = 0.08 = 8\%$

47. Commission = 8% • \$320,000
    = \$25,600

49. Commission rate
    $= \frac{\$66}{\$330} = 0.2 = 20\%$

51. Discount = 25% • \$2520
    = 0.25 • \$2520 = \$630
    Sale price = \$2520 - \$630
    = \$1890

53. Amount of increase
    = 98 cents - 85 cents
    = 13 cents
    Percent increase
    $= \frac{13 \cdot 100}{85} \approx 15.3$

55. Amount of decrease
$\$20,000 - \$18,000 = \$2,000$
Percent decrease
$= \dfrac{2000 \cdot 100}{20,000} = 10$

57. rate of tax

59. rate of discount

61. Sales tax = 8.5% • \$7845
= 0.085 • \$7845
≈ \$666.83

63. Total earned
= \$1280 + 0.035 • \$30,400
= \$2344

65. $\dfrac{333 \text{ cm}}{9 \text{ seconds}} = 37 \dfrac{\text{cm}}{\text{second}}$

67. $\dfrac{\$1.44}{18 \text{ bars}} = \dfrac{144 \text{ cents}}{18 \text{ bars}}$
$= 8 \dfrac{\text{cents}}{\text{bar}}$

69. R and W
$\dfrac{\$2.75}{5 \text{ pounds}} = 0.55 \dfrac{\text{dollars}}{\text{pound}}$
Eastern Family
$\dfrac{\$4.73}{10 \text{ pounds}} = 0.473 \dfrac{\text{dollars}}{\text{pound}}$
Eastern Family sugar has the lower unit price.

71. Oroil
$\dfrac{\$3.95}{5 \text{ quarts}} = 0.79 \dfrac{\text{dollars}}{\text{quart}}$
Lakestate
$\dfrac{\$6.50}{7 \text{ quarts}} \approx 0.93 \dfrac{\text{dollars}}{\text{quart}}$
Oroil has the lower unit price.

Problem Set 6.6

1. *I* = \$400 • 9% • 1
= \$400 • 0.09 = \$36

3. *I* = \$5000 • 8% • 3
= \$5000 • 0.08 • 3
= \$1200

5. *I* = \$860 • 3.7% • 4
= \$860 • 0.037 • 4
= \$127.28

7. *I* = \$440 • 7% • 0.5 = \$440 • 0.07 • 0.5 = \$15.40

9. *I* = \$84 • 1.5% • 3 = \$84 • 0.015 • 3 = \$3.78

11. First year interest = \$600 • 9% • 1 = \$600 • 0.09 = \$54
New principal after first year = \$600 + \$54 = \$654
Second year interest = \$654 • 9% • 1 = \$654 • 0.09 = \$58.86. Amount in account after two years = \$654 + \$58.86 = \$712.86.

13. First year interest = \$4000 • 6.9% • 1 = \$4000 • 0.069 = \$276. New principal after first year = \$4000 + \$276 = \$4276. Second year interest = \$4276 • 6.9% • 1 = \$4276 • 0.069 ≈ \$295.04. Amount in account after two years = \$4276 + \$295.04 = \$4571.04.

15. First half-year interest = \$200 • 15% • 0.5 = \$200 • 0.015 • 0.5 = \$15. New principal after first half-year = \$200 + \$15 = \$215. Second half-year interest = \$215 • 15% • 0.5 = \$215 • 0.15 • 0.5 ≈ \$16.13. Amount in account after one year = \$215 + \$16.13 = \$231.13.

17. First half-year interest = \$9000 • 6.3% • 0.5 = \$9000 • 0.063 • 0.5 = \$283.50. New principal after first half-year = \$9000 + \$283.50 = \$9283.50. Second half-year interest = \$9283.50 • 0.063 • 0.5 ≈ \$292.43. Amount in account after one year = \$9283.50 + \$292.43 = \$9575.93.

19. Interest = %5000 • 8% • 2 = \$5000 • 0.08 • 2 = \$800. Amount in account = \$5000 + 800 = \$5800. Charles earned \$800 interest and had \$5800 after two years.

Problem Set 6.6

21. Interest = \$55,000 • 12% • 4 = \$55,000 • 0.12 • 4 = \$26,400. Total = \$55,000 + \$26,400 = \$81,400. Genotech owes \$26,400 interest on the loan and must pay pack \$81,400.

23. First half-year interest = \$10,000 • 0.06 • 0.5 = \$300. New principal after first half-year = \$10,000 + \$300 = \$10,300. Second half-year interest = \$10,300 • 0.06 • 0.5 = \$309. Amount in account after one year = \$10,300 + \$309 = \$10,609. Jeanelle has \$10,609 in her account after one year.

25. First year interest = \$12,000 • 0.06 • 1 = \$720. New principal after first year = \$12,000 + \$720 = \$12,720. Second year interest = \$12,720 • 0.06 • 1 = \$763.20. Amount in account after two years = \$12,720 + \$763.20 = \$13,483.20. Dave has \$13,483.20 in his account after two years.

27. principal

29. compound

31. First half-year interest = \$3783 • 0.094 • 0.5 ≈ \$177.80. Second half-year interest = (\$3783 + \$177.80) • 0.094 • 0.5 = \$3960.80 • 0.094 • 0.5 ≈ \$186.16. Amount in account after one year = \$3960.80 + \$186.16 = \$4146.96. Joel had \$4146.96 in his account.

33. First year interest = \$7500 • 0.104 • 1 = \$780. Second year interest = (\$7500 + \$780) • 0.104 • 1 = \$8280 • 0.104 = \$861.12. Amount in account after two years = \$8280 + 861.12 = \$9141.12. Barbara had \$9141.12 in her account.

35. $\frac{54}{27} = 2$

37. $\frac{3\frac{1}{8}}{25} = \frac{\frac{25}{8}}{25}$

$\frac{25}{8} \div \frac{25}{1} = \frac{25}{8} \cdot \frac{1}{25} = \frac{1}{8}$

39. $\frac{8}{11}$

41. $\frac{24}{64} = \frac{3}{8}$

43. $\frac{3\frac{2}{3}}{6\frac{1}{9}} = \frac{\frac{11}{3}}{\frac{55}{9}} = \frac{11}{3} \div \frac{55}{9}$

$= \frac{11}{3} \cdot \frac{9}{55}$

$= \frac{\cancel{11}}{\cancel{3}} \cdot \frac{\cancel{3} \cdot 3}{5 \cdot \cancel{11}} = \frac{3}{5}$

Chapter 6 Additional Exercises

1. 42% = 0.42

3. 0.6% = 0.006

5. 0.38 = 38%

7. 0.5 = 50%

9. 71% < 7.1, since 0.71 < 7.1

11. $\frac{9}{10} = 0.9 = 90\%$

$$\begin{array}{r} 0.9 \\ 10\overline{)9.0} \\ \underline{9\ 0} \\ 0 \end{array}$$

13. $\frac{7}{15} \approx 0.467 = 46.7\%$

$$\begin{array}{r} 0.4666 \\ 15\overline{)7.0000} \\ \underline{6\ 0} \\ 1\ 00 \\ \underline{90} \\ 100 \\ \underline{90} \\ 100 \\ \underline{90} \\ 10 \end{array}$$

15. $21\% = \frac{21}{100}$

Chapter 6 Additional Exercises

17. $9\frac{1}{3}\% = \frac{28}{3} \bullet \frac{1}{100} = \frac{7}{75}$

19. $171\frac{1}{2}\% = \frac{343}{2} \bullet \frac{1}{100} = \frac{343}{200}$

21. $\frac{6}{b} = \frac{45}{100}$ $\quad 6 \bullet 100 = 45 \bullet b$
$\frac{600}{45} = b$ $\quad 13.3 \approx b$

23. $\frac{17}{51} = \frac{p}{100}$ $\quad 17 \bullet 100 = 51 \bullet p$
$\frac{1700}{51} = p$ $\quad 33.3 \approx p$

25. $\frac{97.823}{108.645} = \frac{p}{100}$
$97.823 \bullet 100 = 108.645 \bullet p$
$\frac{9782.3}{108.645} = p$
$90.0 \approx p$

27. $\frac{12}{b} = \frac{3}{100}$ $\quad 12 \bullet 100 = 3 \bullet b$
$\frac{1200}{3} = b$ $\quad 400 = b$

29. $\frac{14}{49} = \frac{p}{100}$ $\quad 14 \bullet 100 = 49 \bullet p$
$\frac{1400}{49} = p$ $\quad 28.6 \approx p$

31. $\frac{(200-30)}{200} = \frac{p}{100}$
$170 \bullet 100 = 200 \bullet p$
$\frac{17000}{200} = p$ $\quad 85 = p$
David answered 85% of the questions correctly.

33. $\frac{22}{b} = \frac{8}{100}$ $\quad 22 \bullet 100 = 8 \bullet b$
$\frac{2200}{8} = b$ $\quad 275 = b$
The shipment contained 275 pounds of oranges.

35. $15 = 6\% \bullet b$ $\quad 15 = 0.06 \bullet b$
$\frac{15}{0.06} = b$ $\quad 250 = b$

37. $a = 85\% \bullet 60$
$a = 0.85 \bullet 60$
$a = 51$

39. $64.5 = p\% \bullet 16$;
$64.5 = p \bullet 0.01 \bullet 16$;
$64.5 = p \bullet 0.16$; $\frac{64.5}{0.16} = p$;
$403.1 \approx p$

41. $a = 9\% \bullet 18.50$;
$a = 0.09 \bullet 18.5$; $a \approx 1.7$

43. $a = 16\% \bullet 14$; $a = 0.16 \bullet 14$;
$a = 2.24$; Tax = \$2.24;
Total price =
\$14 + \$2.24 = \$16.24

45. $16 = 32\% \bullet b$; $16 = 0.32 \bullet b$;
$\frac{16}{0.32} = b$; $50 = b$
Snow's Appliance sold 50 small appliances in September.

47. Tax on motorcycle = 8.5% • \$3729 = 0.085 • \$3729 ≈ \$316.97. Total price of motorcycle = \$3729 + \$316.97 = \$4045.97.

49. Kevin's commission = 7% • \$240,000 = 0.07 • \$240,000 = \$16,800.

51. Discount = 35% • \$525 = 0.35 • \$525 = \$183.75. Sale price = \$525 - \$183.75 = \$341.25.

53. Amount of increase = 40 - 32 = 8; Percent increase $= \frac{8 \bullet 100}{32} = 25$.
The class increased by 25%.

55. Amount of decrease = \$6.50 - \$5.50 = \$1.00; Percent decrease $= \frac{1 \bullet 100}{6.50} \approx 15.4$.
Kathy's wages decrease by about 15.4%.

57. Interest = \$6000 • 12% • 2 = \$6000 • 0.12 • 2 = \$1440.

59. Interest = \$400 • 18% • 3 = \$400 • 0.18 • 3 = \$216.

Chapter 6 Additional Exercises

61. First year interest = \$200 • 4% • 1 = \$200 • 0.04 = \$8. Amount in account after one year = \$200 + \$8 = \$208 Second year interest = \$208 • 0.04 • 1 = \$8.32. Amount in account after two years = \$208 + \$8.32 = \$216.32.

63. First half-year interest = \$300 • 14% • 0.5 = \$300 • 0.14 = \$21. Amount in account after first half-year = \$300 + \$21 = \$321 Second half-year interest = \$321 • 0.14 • 0.5 = \$22.47. Amount in account after first year = \$321 + \$22.47 = \$343.47.

65. First quarter-year interest = \$630 • 0.052 • 0.25 = \$8.19. Amount in account after first quarter-year = \$630 + \$8.19 = \$638.19 Second quarter-year interest = \$638.19 • 0.052 • 0.25 ≈ \$8.30. Amount after second quarter-year = \$638.19 + \$8.30 = \$646.49. Third quarter-year interest = \$646.49 • 0.052 • 0.25 ≈ \$8.40. Amount after third quarter-year = \$646.49 + \$8.40 = \$654.89. Fourth quarter-year interest = \$654.89 • 0.052 • 0.25 ≈ \$8.51. Amount in account after one year = \$654.89 + \$8.51 = \$663.40.

67. Ben's interest = \$1800 • 0.015 • 3 = \$81.

69. First half-year interest = \$4000 • 0.14 • 0.5 = \$280. Amount in account after first half-year = \$4000 + \$280 = \$4280. Second half-year interest = \$4280 • 0.14 • 0.5 = \$299.60. Total interest earned in one year = \$280 + \$299.60 = \$579.60. Lelani earned \$579.60 interest in one year.

Chapter 6 Practice Test

1. 37% = 0.37

3. $\frac{3}{5} = 0.6 = 60\%$

5. $\frac{a}{72} = \frac{30}{100}$ $\quad 100 \cdot a = 72 \cdot 30$

   $a = \frac{2160}{100}$ $\quad a = 21.6$

7. Sales tax
   = 0.082 • \$637
   ≈ \$52.23
   Total price
   = \$637 + \$52.23
   = \$689.23
   The sofa costs \$689.23

9. Discount
   = 0.25 • \$170
   = \$42.50
   Sale price
   = \$170 - \$42.50
   = \$127.50
   The silverware is on sale for \$127.50

11. Amount of decrease
   = \$24,000 - \$20,000
   = \$2000
   Percent decrease
   $= \frac{2000 \cdot 100}{24,000} \approx 8.3$
   Jessica's salary decreased by about 8.3%

13. Interest
   = \$1200 • 0.08 • 0.75
   = \$72

Chapter 5 and 6 Cumulative Review

1. $\frac{16}{24} = \frac{2}{3}$

3. $\frac{54 \text{ miles}}{45 \text{ minutes}} = 1\frac{1}{5} \frac{\text{miles}}{\text{minute}}$

5. $\frac{15}{38} \neq \frac{45}{74}$,
   since $15 \cdot 74 \neq 38 \cdot 45$,
   $1110 \neq 1710$

7. $\frac{6.3}{17.64} = \frac{1.4}{x}$
   $6.3 \cdot x = 17.64 \cdot 1.4$
   $x = \frac{24.696}{6.3}$ $\quad x = 3.92$

9. 58.2% = 0.582

Chapter 5 and 6 Cumulative Review

11. $8\% = \frac{8}{100} = \frac{2}{25}$

13. $38 = p\% \bullet 95$
$38 = p \bullet 0.01 \bullet 95$
$38 = p \bullet 0.95$
$\frac{38}{0.95} = p \quad 40 = p$

15. $\frac{(64{,}000 - 24{,}000)}{64{,}000} = \frac{p}{100}$
$40{,}000 \bullet 100 = 64{,}000 \bullet p$
$\frac{4{,}000{,}000}{64{,}000} = p \quad 62.5 = p$
62.5% of Tracy's income is from royalties.

17. $84 = 70\% \bullet b$
$84 = 0.70 \bullet b$
$\frac{84}{0.70} = b$
$120 = b$
Jim's Auto sold 120 cars in November.

19. $c = 0.07 \bullet \$220{,}000$
$= \$15{,}4000$
Steve's commission was $15,400.

21. Increase
$= \$44{,}000 - \$40{,}000$
$= \$4000$
Percent $= \frac{4000 \bullet 100}{40{,}000} = 10$
Daniel's salary increased 10%.

23. Interest
$= \$2400 \bullet 0.07 \bullet 0.5$
$= \$84$

Problem Set 7.1

1. The exponent is 2 and the base is 6.

3. The exponent is 3 and the base is 7.

5. The exponent is 7 and the base is 10.

7. The exponent is 1 and the base is 9.

9. $4 \bullet 4 \bullet 4 = 4^3$

11. $9 \bullet 9 \bullet 9 \bullet 9 = 9^4$

13. $7 \times 7 = 7^2$

15. $2 \times 2 \times 2 \times 2 \times 2 \times 2 = 2^6$

17. $10 \times 10 \times 10 \times 10 = 10^4$

19. $\frac{1}{2} \bullet \frac{1}{2} = \left(\frac{1}{2}\right)^2$

21. $3^2 = 3 \bullet 3 = 9$

23. $2^3 = 2 \bullet 2 \bullet 2 = 8$

25. $12^2 = 12 \bullet 12 = 144$

27. $8^3 = 8 \bullet 8 \bullet 8 = 512$

29. $10^2 = 10 \bullet 10 = 100$

31. $\left(\frac{1}{3}\right)^2 = \frac{1}{3} \bullet \frac{1}{3} = \frac{1}{9}$

33. $1^3 = 1 \bullet 1 \bullet 1 = 1$

35. $(1.4)^2 = 1.4 \times 1.4 = 1.96$

37. $9^2 = 9 \bullet 9 = 81$

39. $\left(\frac{5}{4}\right)^2 = \frac{5}{4} \bullet \frac{5}{4} = \frac{25}{16}$

41. $4 = 4^1$

43. $9 = 9^1$

45. $6^1 = 6$

47. $3^1 = 3$

49. $7^0 = 1$

51. $8^0 = 1$

53. $\left(\frac{3}{5}\right)^0 = 1$

55. $10^1 = 10$

57. $(3.1)^0 = 1$

59. $\left(\frac{11}{4}\right)^0 = 1$

Problem Set 7.1

61. base, exponent

63. cubed, power

65. $7^4 = 2401$

67. $18^3 = 5832$

69. $(9.825)^0 = 1$

71. $\left(\frac{44}{49}\right)^1 \times \left(\frac{971}{369}\right)^0 = \frac{44}{49}$

73. 3^4 is larger than 4^3, since 81 is larger than 64

75. 72.3% = 0.723

77. 0.4% = 0.004

79. 0.482 = 48.2%

81. 8 = 800%

83. 0.06 = 6%

Problem Set 7.2

1. $8^2 = 8 \bullet 8 = 64$

3. $16^2 = 16 \bullet 16 = 256$

5. $10^2 = 10 \bullet 10 = 100$

7. $\left(\frac{1}{2}\right)^2 = \frac{1}{2} \bullet \frac{1}{2} = \frac{1}{4}$

9. $\left(\frac{8}{5}\right)^2 = \frac{8}{5} \bullet \frac{8}{5} = \frac{64}{25}$

11. $\left(\frac{3}{14}\right)^2 = \frac{3}{14} \bullet \frac{3}{14} = \frac{9}{196}$

13. $42^2 = 42 \bullet 42 = 1764$

15. $\left(\frac{7}{30}\right)^2 = \frac{7}{30} \bullet \frac{7}{30} = \frac{49}{900}$

17. $\sqrt{81} = 9$

19. $\sqrt{529} = 23$

21. $\sqrt{441} = 21$

23. $\sqrt{1024} = 32$

25. $\sqrt{6400} = 80$

27. $\sqrt{10,000} = 100$

29. $\sqrt{\frac{16}{25}} = \frac{4}{5}$

31. $\sqrt{\frac{400}{169}} = \frac{20}{13}$

33. $\sqrt{\frac{900}{4}} = \frac{30}{2} = 15$

35. $\sqrt{\frac{1}{4}} = \frac{1}{2}$

37. $\sqrt{\frac{121}{100}} = \frac{11}{10}$

39. $\sqrt{\frac{2500}{1600}} = \frac{50}{40} = \frac{5}{4}$

41. $\sqrt{15} \approx 3.873$

43. $\sqrt{7} \approx 2.646$

45. $\sqrt{11} \approx 3.317$

47. $\sqrt{22} \approx 4.690$

49. $\sqrt{109} \approx 10.440$

51. $\sqrt{200} \approx 14.142$

53. $\sqrt{45} \approx 6.708$

55. $\sqrt{154} \approx 12.410$

57. $1.22 \times \sqrt{4} = 1.22 \times 2 = 2.44$
A person can see 2.44 miles.

59. identical

61. $\sqrt{600} \approx 24.49$

63. $\sqrt{74.8} \approx 8.65$

65. $\sqrt{6.7} \approx 2.59$

67. $\sqrt{0.487} \approx 0.70$

69. $\sqrt{\frac{3}{75}} = \sqrt{0.04} = 0.2$

Problem Set 7.2

71. $\sqrt{\frac{512}{162}} = \sqrt{\frac{256}{81}} = \frac{16}{9}$

73. $4 + 27 \div 3 = 4 + 9 = 13$

75. $(72 - 31) - 10 = 41 - 10 = 31$

77. $\frac{5}{11} \approx 0.455 \approx 45.5\%$

79. $85\% = \frac{85}{100} = \frac{17}{20}$

81. $3.6\% = \frac{36}{1000} = \frac{9}{250}$

Problem Set 7.3

1. $3 \bullet 4^2 + 6 = 3 \bullet 16 + 6$
$= 48 + 6$
$= 54$

3. $70 - 3^2 \bullet 2 = 70 - 9 \bullet 2$
$= 70 - 18$
$= 52$

5. $7^2 - (15 - 6) + 5^2 = 7^2 - 9 + 5^2$
$= 49 - 9 + 25$
$= 40 + 25$
$= 65$

7. $10^2 - 2 \bullet 41 + 12^2 \div 3$
$= 100 - 2 \bullet 41 + 144 \div 3$
$= 100 - 82 + 144 \div 3$
$= 100 - 82 + 48$
$= 18 + 48$
$= 66$

9. $6^2 \bullet 4 \div 2^3 = 36 \bullet 4 \div 8$
$= 144 \div 8 = 18$

11. $14^2 + \frac{0}{6} = 196 + \frac{0}{6}$
$= 196 + 0$
$= 196$

13. $\left(\frac{2}{3}\right)^2 \bullet \frac{1}{2} + \frac{3}{8} = \frac{4}{9} \bullet \frac{1}{2} + \frac{3}{8}$
$= \frac{4}{18} + \frac{3}{8}$
$= \frac{4}{18} \bullet \frac{4}{4} + \frac{3}{8} \bullet \frac{9}{9}$
$= \frac{16}{72} + \frac{27}{72} = \frac{43}{72}$

15. $\left(\frac{8}{5} - \frac{7}{10}\right)^2 - \frac{3}{5} + \left(\frac{1}{2}\right)^2$
$= \left(\frac{16}{10} - \frac{7}{10}\right)^2 - \frac{3}{5} + \left(\frac{1}{2}\right)^2$
$= \left(\frac{9}{10}\right)^2 - \frac{3}{5} + \left(\frac{1}{2}\right)^2$
$= \frac{81}{100} - \frac{3}{5} + \frac{1}{4}$
$= \frac{81}{100} - \frac{60}{100} + \frac{25}{100}$
$= \frac{21}{100} + \frac{25}{100}$
$= \frac{46}{100}$
$= \frac{23}{50}$

17. $15 + 3 \bullet (1.2)^2 = 15 + 3 \bullet 1.44$
$= 15 + 4.32$
$= 19.32$

19. $1.37 \times 10^4 - 4.28 \times 10^3$
$= 1.37 \times 10{,}000 - 4.28 \times 1000$
$= 13{,}700 - 4.28 \times 1000$
$= 13{,}700 - 4280$
$= 9420$

21. $16 \div (0.01)^3 - 40 \div (0.4)^3$
$= 16 \div 0.0001 - 40 \div 0.064$
$= 160{,}000 - 40 \div 0.064$
$= 160{,}000 - 625$
$= 159{,}375$

23. $(6.4)^2 - 11.4 \times 7.5 \div 3.75$
$= 40.96 - 11.4 \times 7.5 \div 3.75$
$= 40.96 - 85.5 \div 3.75$
$= 40.96 - 22.8$
$= 18.16$

25. $(5.28)^3 - (8.27 - 4.156) + (1.5)^3 \times 80$
$= (5.28)^3 - 4.114 + (1.5)^3 \times 80$
$= 147.19795 - 4.114 + 3.375 \times 80$
$= 147.19795 - 4.114 + 270$
$= 143.08395 + 270$
$= 413.08395$

27. $3 \times \{(100 - 40 \div 4) + [(45 \div 9)^2 - 8]\}$
$= 3 \times [(100 - 10)\ (5^2 - 8)]$
$= 3 \times [(100 - 10) + (25 - 8)]$
$= 3 \times (90 + 17)$
$= 3 \times 107$
$= 321$

Problem Set 7.3

29. Does $5 \times 18 = 6 \cdot 15$?
Since $90 = 90$, the proportion is true.

31. Does $17 \cdot 49 = 16 \cdot 51$?
Since $833 \neq 816$, the proportion is false.

33. $a = 25\% \cdot 47$
$a = 0.25 \cdot 47$
$a = 11.75$
$a \approx 11.8$

35. $32 = 8\% \cdot b$
$32 = 0.08 \cdot b$
$\frac{32}{0.08} = b$
$400 = b$

37. $15 = p\% \cdot 76$
$15 = p \cdot 0.01 \cdot 76$
$15 = p \cdot 0.76$
$\frac{15}{0.76} = p$
$19.7 \approx p$

Chapter 7 Additional Exercises

1. $8 \times 8 \times 8 = 8^3$

3. $10 \times 10 \times 10 \times 10 = 10^4$

5. $8^2 = 8 \cdot 8 = 64$

7. $5^0 = 1$

9. $7^1 = 7$

11. $\frac{37}{51} \times \left(\frac{2784}{9826}\right)^0 \times \left(\frac{1}{2}\right)^3$
$= \frac{37}{51} \times 1 \times \frac{1}{8} = \frac{37}{408}$

13. $48^2 = 2304$

15. $\left(\frac{11}{20}\right)^2 = \frac{121}{400}$

17. $\sqrt{961} = 31$

19. $\sqrt{\frac{1600}{225}} = \frac{40}{15} = \frac{8}{3}$

21. $\sqrt{1\frac{21}{100}} = \sqrt{\frac{121}{100}} = \frac{11}{10}$

23. $\sqrt{111} \approx 10.536$

25. Inventory
$= 5 \cdot \sqrt{16} = 5 \cdot 4 = 20$
The store's optimal inventory is 20 dishwashers

27. $3^4 \div 3^2 \cdot 16 = 81 \div 9 \cdot 16$
$= 9 \cdot 16$
$= 144$

29. $(38-34)^2+7^2 = 4^2+7^2$
$= 16+49$
$= 65$

31. $4.28 \times 10^3-16^2$
$= 4.28 \times 1000-256$
$= 4280-256$
$= 4024$

Chapter 7 Practice Test

1. $7 \cdot 7 \cdot 7 \cdot 7 = 7^4$

3. $2^3 = 2 \cdot 2 \cdot 2 = 8$

5. $19^0 = 1$

7. $\left(\frac{9}{8}\right)^2 = \frac{81}{64}$

9. $\sqrt{225} = 15$

11. $\sqrt{15} \approx 3.873$

13. $8^2-9 \cdot 5 = 64-9 \cdot 5$
$= 64-45$
$= 19$

Problem Set 8.1

1. $\frac{1 \text{ \sout{pound}}}{1} \cdot \frac{16 \text{ ounces}}{1 \text{ \sout{pound}}}$
$= 16$ ounces

3. $\frac{1 \text{ \sout{quart}}}{1} \cdot \frac{2 \text{ pints}}{1 \text{ \sout{quart}}} = 2$ pints

5. $\frac{1 \text{ \sout{mile}}}{1} \cdot \frac{5280 \text{ feet}}{1 \text{ \sout{mile}}}$
$= 5280$ feet

7. $\frac{8 \text{ \sout{pints}}}{1} \cdot \frac{2 \text{ cups}}{1 \text{ \sout{pint}}} = 16$ cups

9. $\frac{9 \text{ \sout{gallons}}}{1} \cdot \frac{4 \text{ quarts}}{1 \text{ \sout{gallon}}}$
$= 36$ quarts

Problem Set 8.1

11. $\frac{7 \text{ tons}}{1} \cdot \frac{2000 \text{ pounds}}{1 \text{ ton}} = 14{,}000 \text{ pounds}$

13. $\frac{8\frac{1}{4} \text{ days}}{1} \cdot \frac{24 \text{ hours}}{1 \text{ day}} = 198 \text{ hours}$

15. $\frac{11 \text{ feet}}{1} \cdot \frac{12 \text{ inches}}{1 \text{ foot}} = 132 \text{ inches}$

17. $\frac{12 \text{ cups}}{1} \cdot \frac{1 \text{ pint}}{2 \text{ cups}} = 6 \text{ pints}$

19. $\frac{48 \text{ hours}}{1} \cdot \frac{1 \text{ day}}{24 \text{ hours}} = 2 \text{ days}$

21. $\frac{15{,}840 \text{ feet}}{1} \cdot \frac{1 \text{ mile}}{5280 \text{ feet}} = 3 \text{ miles}$

23. $\frac{92 \text{ inches}}{1} \cdot \frac{1 \text{ foot}}{12 \text{ inches}} = \frac{23}{3} \text{ feet} = 7\frac{2}{3} \text{ feet}$

25. $\frac{5\frac{1}{2} \text{ quarts}}{1} \cdot \frac{1 \text{ gallon}}{4 \text{ quarts}} = \frac{11}{8} \text{ gallons} = 1\frac{3}{8} \text{ gallons}$

27. $\frac{1 \text{ yard}}{1} \cdot \frac{3 \text{ feet}}{1 \text{ yard}} = 3 \text{ feet}$

29. $\frac{5 \text{ quarts}}{1} \cdot \frac{2 \text{ pints}}{1 \text{ quart}} = 10 \text{ pints}$

31. $\frac{108 \text{ inches}}{1} \cdot \frac{1 \text{ foot}}{12 \text{ inches}} = 9 \text{ feet}$

33. $\frac{78 \text{ inches}}{1} \cdot \frac{1 \text{ foot}}{12 \text{ inches}} = \frac{13}{2} \text{ feet} = 6\frac{1}{2} \text{ feet}$

35. $\frac{14 \text{ gallons}}{1} \cdot \frac{4 \text{ quarts}}{1 \text{ gallon}} = 56 \text{ quarts}$

37. $\frac{31{,}680 \text{ feet}}{1} \cdot \frac{1 \text{ mile}}{5280 \text{ feet}} = 6 \text{ miles}$

39. $\frac{15 \text{ inches}}{1} \cdot \frac{1 \text{ foot}}{12 \text{ inches}} = \frac{5}{4} \text{ feet} = 1\frac{1}{4} \text{ feet}$

41. $\frac{3\frac{1}{4} \text{ days}}{1} \cdot \frac{24 \text{ hours}}{1 \text{ day}} = 78 \text{ hours}$

43. $\frac{8 \text{ gallons}}{1} \cdot \frac{4 \text{ quarts}}{1 \text{ gallon}} = 32 \text{ quarts}$
Kevin bought 32 quarts of ice cream.

45. $\frac{21 \text{ yards}}{1} \cdot \frac{3 \text{ feet}}{1 \text{ yard}} = 63 \text{ feet}$
The foundation is 63 feet long.

47. $\frac{20 \text{ ounces}}{1} \cdot \frac{1 \text{ pound}}{16 \text{ ounces}} = \frac{5}{4} \text{ pounds} = 1\frac{1}{4} \text{ pounds}$
The rod weighs $1\frac{1}{4}$ pounds.

49. $\frac{3.2 \text{ minutes}}{1} \cdot \frac{60 \text{ seconds}}{1 \text{ minute}} = 192 \text{ seconds}$
Susan ran the race in 192 seconds.

51. $\frac{2\frac{1}{2} \text{ pounds}}{1} \cdot \frac{1 \text{ ton}}{2000 \text{ pounds}} \cdot 1800 = 2\frac{1}{4} \text{ tons}$
A load of 1800 bricks weighs $2\frac{1}{4}$ tons.

53. $\frac{15 \text{ yards}}{1} \cdot \frac{3 \text{ feet}}{1 \text{ yard}} = 45 \text{ feet}$
Julie has 45 feet of fencing, which is not enough. 51 - 45 = 6. Julie needs 6 more feet of fencing to enclose the yard.

55. $\frac{19{,}940 \text{ feet}}{1} \cdot \frac{1 \text{ mile}}{5280 \text{ feet}} \approx 3.8 \text{ miles}$

Problem Set 8.1

57. $\frac{7\ \cancel{yards}}{1} \bullet \frac{3\ \cancel{feet}}{1\ \cancel{yard}} \bullet \frac{12\ \text{inches}}{1\ \cancel{foot}}$
$= 252$ inches

59. $\frac{5\ \cancel{miles}}{1} \bullet \frac{5280\ \cancel{feet}}{1\ \cancel{mile}} \bullet \frac{1\ \text{yard}}{3\ \cancel{feet}}$
$= 8800$ yards

61. $\frac{190{,}080\ \cancel{inches}}{1} \bullet \frac{1\ \cancel{foot}}{12\ \cancel{inches}}$
$\bullet \frac{1\ \text{mile}}{5280\ \cancel{feet}} = 3$ miles

63. $a = 57\% \bullet 94$
$a = 0.57 \bullet 94$
$a = 53.58$
$a \approx 53.6$

65. $49 = p\% \bullet 72$
$49 = p \bullet 0.01 \bullet 72$
$49 = p \bullet 0.72$
$\frac{49}{0.72} = p$
$68.1 \approx p$

67. $36 = 20\% \bullet b$
$36 = 0.20 \bullet b$
$\frac{36}{0.20} = b$
$180 = b$

69. $a = 3\% \bullet 78$
$a = 0.30 \bullet 78$
$a = 2.34$
$a \approx 2.3$

71. $56 = p\% \bullet 38$
$56 = p \bullet 0.01 \bullet 38$
$56 = p \bullet 0.38$
$\frac{56}{0.38} = p$
$147.4 \approx p$

Problem Set 8.2

1. 1 foot + 15 inches
$= 1\ \text{ft} + \frac{15\ \text{in.}}{12\ \text{in. per ft}}$
$= 1$ ft $+ 1$ ft 3 in.
$= 2$ ft 3 in.

3. 1 gallon + 5 quarts
$= 1\ \text{gal} + \frac{5\ \text{qt}}{4\ \text{qt per gal}}$
$= 1$ gal $+ 1$ gal 1 qt
$= 2$ gal 1 qt

5. 1 pound + 22 ounces
$= 1\ \text{lb} + \frac{22\ \text{oz}}{16\ \text{oz per lb}}$
$= 1$ lb $+ 1$ lb 6 oz
$= 2$ lb 6 oz

7. 5 yards + 7 feet
$= 5\ \text{yd} + \frac{7\ \text{ft}}{3\ \text{ft per yd}}$
$= 5$ yd $+ 2$ yd 1 ft
$= 7$ yd 1 ft

9. 1 pints + 5 cups
$= 1\ \text{pt} + \frac{5\ \text{cups}}{2\ \text{cups per pt}}$
$= 1$ pt $+ 2$ pt 1 cups
$= 3$ pt 1 cup

11. 1 ton + 3000 pounds
$= 1\ \text{ton} + \frac{3000\ \text{lb}}{2000\ \text{lb per ton}}$
$= 1$ ton $+ 1$ ton 1000 lb
$= 2$ tons 1000 lb

13. 2 weeks + 9 days + 28 hours
$= 2$ wks $+ 9$ days
$+ \frac{28\ \text{h}}{24\ \text{h per day}}$
$= 2$ wks $+ 9$ days
$+ 1$ day 4 h
$= 2$ wks $+ 10$ days $+ 4$ h
$= 2\ \text{wks} + \frac{10\ \text{days}}{7\ \text{days per wk}} + 4\ \text{h}$
$= 3$ wks 3 days 4 h

15. 4 gallons + 7 quarts + 3 pints
$= 4$ gal $+ 7$ qt
$+ \frac{3\ \text{pt}}{2\ \text{pt per qt}}$
$= 4$ gal $+ 7$ qt
$+ 1$ qt 1 pt
$= 4$ gal $+ 8$ qt $+ 1$ pt
$= 4\ \text{gal} + \frac{8\ \text{qt}}{4\ \text{qt per gal}} + 1\ \text{pt}$
$= 4$ gal $+ 2$ gal $+ 1$ pt
$= 6$ gal 1 pt

17. 12 yards + 4 feet + 16 inches
$= 12$ yd $+ 4$ ft
$+ \frac{16\ \text{in.}}{12\ \text{in. per ft}}$
$= 12$ yd $+ 4$ ft
$+ 1$ ft 4 in.
$= 12\ \text{yd} + \frac{5\ \text{ft}}{3\ \text{ft per yd}} + 4\ \text{in.}$
$= 12$ yd $+ 1$ yd 2 ft $+ 4$ in.
$= 13$ yd 2 ft 4 in.

Problem Set 8.2

19.
7 gallons 3 quarts
+ 4 gallons 2 quarts
11 gallons 5 quarts
= 11 gal + 1 gal 1 qt
= 12 gal 1 qt

21.
8 feet 4 inches
+ 2 feet 17 inches
10 feet 21 inches
= 10 ft + 1 ft 9 in.
= 11 ft 9 in.

23.
1 hour 52 minutes
+ 8 hours 19 minutes
9 hours 71 minutes
= 9 h + 1 h 11 min
= 10 h 11 min

25.
4 quarts 3 pints 1 cup
+ 3 quarts 5 pints 9 cups
7 quarts 8 pints 10 cups
= 7 qt + 8 pt + 5 pt
= 7 qt + 13 pt
= 7 qt + 6 qt 1 pt
= 13 qt 1 pt

27.
9 feet 11 inches
− 6 feet 7 inches
3 feet 4 inches

29.
23 25
~~24~~ pounds ~~9~~ ounces
− 16 pounds 14 ounces
7 pounds 11 ounces

31.
8 39
~~9~~ days ~~15~~ hours
− 3 days 22 hours
5 days 17 hours

33.
6 4
5 gallons ~~7~~ quarts ~~2~~ pints
− 2 gallons 4 quarts 3 pints
3 gallons 2 quarts 1 pint

35.
9 pounds 12 ounces
× 3
27 pounds 36 ounces
= 27 lb + 2 lb 4 oz
= 29 lb 4 oz

37.
7 feet 5 inches
× 7
49 feet 35 inches
= 49 ft + 2 ft 11 in.
= 51 ft 11 in.

39.
4 tons 300 pounds
× 9
36 tons 2700 pounds
= 36 tons + 1 ton 700 lb
= 37 tons 700 lb

41.
5 wks 7 days 4 h
× 7
35 wks 49 days 28 h
= 35 wks + 49 days + 1 day 4 h
= 35 wks + 50 days + 4 h
= 35 wks + 7 wks 1 day + 4 h
= 42 wks 1 day 4 h

43.
3 yards 4 feet
2)6 yards 8 feet
6 yards
0 8 feet
8 feet
0

45.
2 feet 7 inches
3)7 feet 9 inches
6 feet
1 foot = 12 inches
21 inches
21 inches
0

47.
2 miles 85 feet
9)18 miles 765 feet
18 miles
0 765 feet
765 feet
0

49.
2 h 17 min 5 s
4)9 h 8 min 20 s
8 h
1 h = 60 min
68 min
68 min
0 20 s
20 s
0

51.
9 gallons + 8 quarts
$= 9 \text{ gal} + \dfrac{8 \text{ qt}}{4 \text{ qt per gal}}$
= 9 gal + 2 gal
= 11 gal
Keith bought 11 gallons of milk.

Problem Set 8.2

53.
2 feet 5 inches
× 3
6 feet 15 inches
= 6 ft + 1 ft 3 in.
= 7 ft 3 in.
The board should be 7 feet 3 inches.

55.
11 feet 2 inches
+ 9 feet 11 inches
20 feet 13 inches
= 20 ft + 1 ft 1 in.
= 21 ft 1 in.
The room is 21 feet 1 inch long.

57.
1 h 37 min
4)6 h 28 min
4 h
2 h = 120 min
148 min
148 min
0
The new machine completed the job in 1 hour 37 minutes.

59.
9.2 gallons 3.7 quarts
+ 7.9 gallons 4.8 quarts
17.1 gallons 8.5 quarts
= 17.1 gal
+2 gal 0.5 qt
= 19.1 gal 0.5 qt

61.
59.2 feet 11.8 inches
+27.3 feet 9.3 inches
31.9 feet 2.5 inches

63.
1 h 25 min 24 s
7)9 h 57 min 48 s
7 h
2 h = 120 min
177 min
175 min
2 min 120 s
168 s
168 s
0

65. Sales tax = 5% • \$325
= 0.05 • \$325 = \$16.25

67. Commission = 7% • \$3240
= 0.07 • \$3240
= \$226.80

69. Discount = 20% • \$74.20
= 0.20 • \$74.20
≈ \$14.84
Sale price = \$74.20 - \$14.84
= \$59.36

71. Amount of increase
= \$34,000 - \$30,000
= \$4000
Percent increase
$= \dfrac{4000 \cdot 100}{30{,}000} \approx 13.3$
Linda's salary increased by about 13.3%.

73. Amount of decrease
= \$500 - \$475 = \$25
Percent decrease
$= \dfrac{25 \cdot 100}{500} = 5$
Deborah's rent decreased by 5%.

Problem Set 8.3

1. 1 km = 1000 m

3. 1 hm = 100 m

5. 1 dam = 10 m

7. 1 m = 10 dm

9. 1 m = 100 cm

11. 1 m = 1000 mm

13. 9 km = 9000 m

15. 53.7 m = 5370 cm

17. 4203 m = 4,203,000 mm

19. 73.4 cm = 0.734 m

21. 9347 m = 9.347 km

23. 1 mm = 0.1 cm

25. 15 dm = 1.5 m

27. 17 dam = 170 m

29. 8 km = 800,000 cm

31. 764 m = 764,000 mm

33. 9254 mm = 0.009254 km

Problem Set 8.3

35. 8.2 mL = 0.0082 L

37. 98.1 L = 98,100 mL

39. 1 kg = 1000 g

41. 1 cg = 0.01 g

43. 1 mg = 0.001 g

45. 14 kg = 14,000 g

47. 783 g = 0.783 kg

49. 53 cg = 0.53 g

51. 6.8 g = 680 cg

53. 5.4 kg = 540,000 cg

55. 90 mg = 9 cg

57. 15 dag = 0.15 kg

59. 70,489 mg = 0.070489 kg

61. 183 cm = 1830 mm
John's height is 1830 mm.

63. 0.625 mg = 0.0625 cg
The dose is 0.0625 cg.

65. 2.4 L = 2400 mL
$\frac{5 \cdot 2400\text{ mL}}{200\text{ mL}} = \frac{12{,}000\text{ mL}}{200\text{ mL}} = 60$
Juana can serve 60 guests.

67. 15 kg = 15,000 g
$\frac{15{,}000\text{ g}}{300\text{ g}} = 50$
The store gets 50 packages from the container.

69. meter

71. gram

73. $200\frac{\text{mg}}{\text{day}} \cdot 6\text{ weeks} \cdot \frac{7\text{ days}}{1\text{ week}}$
$\cdot\, 4 = 33{,}600\text{ mg}$
33,600 mg = 33.6 g
Jeff takes 33.6 g of vitamin C.

75. Interest = \$8000 • 4% • 5
= \$8000 • 0.04 • 5
= \$1600

77. Interest = \$5000 • 11.2% • 3
= \$5000 • 0.112 • 3
= \$1680

79. Interest = \$740 • 11% • $\frac{3}{4}$
= \$740 • 0.11 • 0.75
= \$61.05

81. First year interest = \$6000 • 0.08 • 1 = \$480. Amount in account after first year = \$6000 + \$480 = \$6480. Second year interest = \$6480 • 0.08 • 1 = \$518.40. Amount in account after two years = \$6480 + \$518.40 = \$6998.40.

83. First year interest = \$4000 • 0.09 • 1 = \$360. Amount in account after first year = \$4000 + \$360 = \$4360. Second year interest = \$4360 • 0.09 • 1 = \$392.40. Amount in account after two years = \$4360 + \$392.40 = \$4752.40.

Problem Set 8.4

1. $7\text{ in.} \approx \frac{7\ \cancel{\text{in.}}}{1} \cdot \frac{2.54\text{ cm}}{1\ \cancel{\text{in.}}}$
$\approx 17.78\text{ cm}$

3. $17.6\text{ lb} \approx \frac{17.6\ \cancel{\text{lb}}}{1} \cdot \frac{1\text{ kg}}{2.2\ \cancel{\text{lb}}}$
$\approx 8\text{ kg}$

5. $48.3\text{ km} \approx \frac{48.3\ \cancel{\text{km}}}{1} \cdot \frac{1\text{ mi}}{1.61\ \cancel{\text{km}}}$
≈ 30 miles

7. 5.83 qt
$\approx \frac{5.83\ \cancel{\text{qt}}}{1} \cdot \frac{1\text{ L}}{1.06\ \cancel{\text{qt}}}$
$\approx 5.5\text{ L}$

9. $197\text{ in.} \approx \frac{197\ \cancel{\text{in.}}}{1} \cdot \frac{1\text{ m}}{39.4\ \cancel{\text{in.}}}$
$\approx 5\text{ m}$

11. $5.9\text{ L} \approx \frac{5.9\ \cancel{\text{L}}}{1} \cdot \frac{1.06\text{ qt}}{1\ \cancel{\text{L}}}$
≈ 6.254 quarts

Problem Set 8.4

13. $30 \text{ lb} \approx \frac{30 \text{ lb}}{1} \bullet \frac{454 \text{ g}}{1 \text{ lb}}$
$\approx 13{,}620 \text{ g}$

15. $F = \frac{9}{5} \bullet 35 + 32 = 63 + 32 = 95$

17. $F = \frac{9}{5} \bullet 84 + 32 = 151.2 + 32$
$= 183.2$

19. $C = \frac{5}{9} \bullet (86 - 32) = \frac{5}{9} \bullet 100$
≈ 30

21. $C = \frac{5}{9} \bullet (132 - 32) = \frac{5}{9} \bullet 100$
≈ 55.6

23. $57 \text{ qt} \approx \frac{57 \text{ qt}}{1} \bullet \frac{1 \text{ L}}{1.06 \text{ qt}}$
$\approx 53.77 \text{ L}$

25. $18 \text{ lb} \approx \frac{18 \text{ lb}}{1} \bullet \frac{1 \text{ kg}}{2.20 \text{ lb}}$
$\approx 8.18 \text{ kg}$

27. $15 \text{ m} \approx \frac{15 \text{ m}}{1} \bullet \frac{39.4 \text{ in.}}{1 \text{ m}}$
≈ 591 inches

29. $7284 \text{ g} \approx \frac{7284 \text{ g}}{1} \bullet \frac{1 \text{ lb}}{454 \text{ g}}$
≈ 16.04

31. $C = \frac{5}{9} \bullet (103 - 32) = \frac{5}{9} \bullet 71$
≈ 39.44
Jan's temperature is about $39.44°$ C.

33. $3.2 \text{ lb} \approx \frac{3.2 \text{ lb}}{1} \bullet \frac{454 \text{ g}}{1 \text{ lb}}$
$\approx 1452.8 \text{ g}$
There are about 1452.8 g of hamburger in 3.2 lbs of hamburger.

35. $27 \text{ in.} \approx \frac{27 \text{ in.}}{1} \bullet \frac{1 \text{ in.}}{2.54 \text{ cm}}$
$\approx 68.58 \text{ cm}$
The ribbon is about 68.58 cm long.

37. $15 \text{ cm} \approx \frac{15 \text{ cm}}{1} \bullet \frac{1 \text{ in}}{2.54 \text{ cm}}$
≈ 5.91 inches
Since 15 cm ≈ 5.91 inches, 15 inches is larger than 15 cm.

39. $789 \text{ g} \approx \frac{789 \text{ g}}{1} \bullet \frac{1 \text{ lb}}{454 \text{ g}} \bullet \frac{16 \text{ ozs}}{1 \text{ lb}} \approx 27.81$ ounc

41. $8 \bullet 8 \bullet 8 = 8^3$

43. $2 \times 2 \times 2 \times 2 = 2^4$

45. $6^2 = 6 \bullet 6 = 36$

47. $15^0 = 1$

49. $4^3 = 4 \bullet 4 \bullet 4 = 64$

Chapter 8 Additional Exercises

1. $9 \text{ ft} = \frac{9 \text{ ft}}{1} \bullet \frac{12 \text{ in.}}{1 \text{ ft}}$
$= 108$ inches

3. $63 \text{ days} = \frac{63 \text{ days}}{1} \bullet \frac{1 \text{ wk}}{7 \text{ day}}$
$= 9$ weeks

5. $2\frac{1}{4} \text{ mi} = \frac{2.25 \text{ mi}}{1} \bullet \frac{5280 \text{ f}}{1 \text{ mi}}$
$= 11{,}880$ feet

7. $3 \text{ gal} = \frac{3 \text{ gal}}{1} \bullet \frac{4 \text{ qt}}{1 \text{ gal}} \bullet \frac{2 \text{ pt}}{1 \text{ qt}} \bullet \frac{2 \text{ cu}}{1 \text{ pt}}$
$= 48$ cups

9. $988 \text{ ft} = \frac{988 \text{ ft}}{1} \bullet \frac{1 \text{ yd}}{3 \text{ ft}}$
$\approx 329\frac{1}{3}$ yards
The length of the fiel $329\frac{1}{3}$ yards.

11. 17 yards + 10 feet + 48 ir
= 17 yd + 10 ft + 4
= 17 yd + 14 ft
= 17 yd + 4 yd 2 f
= 21 yd 2 ft

Chapter 8 Additional Exercises

13.
```
    7 qt 3 pt  9 cups
 + 3 qt 4 pt  1 cup
  10 qt 7 pt 10 cups
         = 10 qt+7 pt+5 pt
         = 10 qt+12 pt
         = 10 qt+6 qt
         = 16 qt
```

15.
```
  8        10
  9 yards  7 feet 4 inches
 -4 yards  9 feet 2 inches
  4 yards  1 feet 2 inches
```

17.
```
   7 h 15 min  20 s
 ×                6
  42 h 90 min 120 s
       = 42 h+90 min+2 min
       = 42 h+92 min
       = 42 h+1 h 32 min
       = 43 h 32 min
```

19.
```
    2 gal 2 qt 2 pt
 4)8 gal 9 qt  6 pt
   8 gal
   0     9 qt
         8 qt
         1 qt = 2 pt
                8 pt
                8 pt
                0
```

21. 72 m = 72,000 mm

23. 725 mm = 72.5 cm

25. 42.5 cg = 0.425 g

27. 1.2 g = 1200 mm

$$1200 \div 150 = 8$$

Christine must take 8 tablets each day.

29. $3\text{ kg} \approx \frac{3\text{ kg}}{1} \bullet \frac{2.2\text{ lb}}{1\text{ kg}} \approx 6.6\text{ lb}$

31. $275.8\text{ m} \approx \frac{275.8\text{ m}}{1} \bullet \frac{39.4\text{ in}}{1\text{ m}} \approx 10{,}866.52\text{ inches}$

33. $12.88\text{ km} \approx \frac{12.88\text{ km}}{1} \bullet \frac{1\text{ mi}}{1.61\text{ km}} \approx 8\text{ miles}$

35. $C = \frac{5}{9} \bullet (143-32) = \frac{5}{9} \bullet 111 \approx 61.7$

Chapter 8 Practice Test

1. $\frac{48\text{ in.}}{1} \bullet \frac{1\text{ ft}}{12\text{ in.}} = 4\text{ ft}$

3. $\frac{5\text{ gal}}{1} \bullet \frac{4\text{ qt}}{1\text{ gal}} = 20\text{ qt}$

5.
```
   5 gal 4 qt 2 pt
 + 3 gal 2 qt 1 pt
   8 gal 6 qt 3 pt
      = 8 gal+6 qt
            +1 qt 1 pt
      = 8 gal+7 qt+1 pt
      = 8 gal+1 gal 3 qt
            +1 pt
      = 9 gal 3 qt 1 pt
```

7.
```
   4 mi  960 ft
 ×            6
  24 mi 5760 ft
      = 24 mi+1 mi 480 ft
      = 25 mi 480 ft
```

9. 4500 m = 4.5 km

11. 457 L = 457,000 mL

13. $9\text{ lb} \approx \frac{9\text{ lb}}{1} \bullet \frac{454\text{ g}}{1\text{ lb}} \approx 4086\text{ g}$

15. $7\text{ L} \approx \frac{7\text{ L}}{1} \bullet \frac{1.06\text{ qt}}{1\text{ L}} = 7.42\text{ quarts}$

17. $F = \frac{9}{5} \bullet 30 + 32 = 54 + 32 = 86$

19. $\frac{8\text{ qts}}{1} \bullet \frac{2\text{ pts}}{1\text{ qt}} \bullet \frac{2\text{ cups}}{1\text{ pt}} = 32\text{ cups}$

Jim made 32 cups of punch.

21. 298 g • 24 = 7152 g
7152 g = 7.152 kg
There are 7.152 kg of soup in 24 cans.

Chapters 7 and 8 Cumulative Review

1. $2 \bullet 2 \bullet 2 = 2^3$

3. $4^3 = 4 \bullet 4 \bullet 4 = 64$

Chapters 7 and 8 Cumulative Review

5. $8^0 = 1$

7. $\left(\frac{11}{12}\right)^2 = \frac{121}{144}$

9. $\sqrt{900} = 30$

11. $\sqrt{84} \approx 9.165$

13. $9^2 - 3 \bullet 2 = 81 - 3 \bullet 2 = 81 - 6 = 75$

15. $\frac{36\ \cancel{ft}}{1} \bullet \frac{1\ yd}{3\ \cancel{ft}} = 12$ yards

17. $\frac{4\frac{1}{4}\ \cancel{lb}}{1} \bullet \frac{16\ oz}{1\ \cancel{lb}} = 68$ ounces

19. $\frac{7\ \cancel{yd}}{1} \bullet \frac{3\ ft}{1\ \cancel{yd}} = 21$ feet

21.
```
   9 yards 2 feet  7 inches
 + 2 yards 1 foot 11 inches
  11 yards 3 feet 18 inches
         = 11 yd + 3 ft + 1 ft 6 in.
         = 11 yd + 4 ft + 6 in.
         = 11 yd + 1 yd 1 ft + 6 in.
         = 12 yd 1 ft 6 in.
```

23.
```
   5 tons 1500 pounds
 ×                  3
  15 tons 4500 pounds
       = 15 tons + 2 tons 500 lb
       = 17 tons 500 lb
```

25.
```
  8   83
  9̸ h 2̸3 min
 -6 h 35 min
  2 h 48 min
```
Laurie worked 2 hours 48 minutes longer on Thursday.

27. 320 m = 32,000 cm

29. 5243 mL = 5.243 L

31. 176 cm - 160 cm = 16 cm
16 cm = 160 mm. The difference in their height is 160 mm.

33. $2270\ g \approx \frac{2770\ \cancel{g}}{1} \bullet \frac{1\ lb}{454\ \cancel{g}} \approx 5\ lb$

35. $15.9\ qt \approx \frac{15.9\ \cancel{qt}}{1} \bullet \frac{1\ L}{1.06\ \cancel{qt}} \approx 15\ L$

37. $F = \frac{9}{5} \bullet 71 + 32 = 127.8 + 32$
$= 159.8$

39. $104\ km \approx \frac{104\ km}{1} \bullet \frac{1\ mi}{1.61\ km}$
≈ 64.60
You can drive 64.6 miles per hour and stay within the speed limit.

Problem Set 9.1

1. $P = 4\ ft + 3\ ft + 6\ ft = 13\ ft$

3. $P = 2 \times 7.2\ cm + 2 \times 4.8\ cm$
$= 24\ cm$

5. $P = \frac{2}{5}\ yd + \frac{2}{5}\ yd + \frac{2}{5}\ yd + \frac{2}{5}\ yd$
$= 4 \bullet \frac{2}{5}\ yd = \frac{8}{5}\ yd$
$= 1\frac{3}{5}\ yd$

7. $P = 2 \bullet 4\ in. + 2 \bullet 9\ in.$
$= 26\ in.$

9. $P = 2 \times 8.3\ ft + 2 \times 9.7\ ft$
$= 36\ ft$

11. $P = 2 \bullet 5\frac{1}{8}\ yd + 2 \bullet 6\frac{3}{4}\ yd$
$= \frac{2}{1} \bullet \frac{41}{8}\ yd$
$+ \frac{2}{1} \bullet \frac{27}{4}\ yd \bullet \frac{2}{2}$
$= \frac{190}{8}\ yd = 23\frac{3}{4}\ yd$

13. $P = 4 \bullet 18\ in. = 72\ in.$

15. $P = 4 \times 7.1\ m = 28.4\ m$

17. $P = 4 \bullet 7\frac{3}{8}\ ft = \frac{4}{1} \bullet \frac{59}{8}\ ft$
$= \frac{236}{8}\ ft = 29\frac{1}{2}\ ft$

19. $P = 2 \bullet 15\ ft + 2 \bullet 12\ ft = 54\ ft$
Lisa needs 54 feet of border.

Problem Set 9.1

21. $P = 2 \cdot 183 \text{ cm} + 2 \cdot 61 \text{ cm}$
$= 488 \text{ cm}$
The perimeter of the table is 488 centimeters.
Cost $= 488 \times 0.06 = 29.28$
The edging will cost $29.28.

23. perimeter

25. same

27. $P = 78.4 \text{ mm} + 97.3 \text{ mm} + 120.8 \text{ mm}$
$= 296.5 \text{ mm}$

29. $P = 2 \times 508.92 \text{ m} + 2 \times 1104.95 \text{ m}$
$= 3227.74 \text{ m}$

31. Fencing $= 25 \text{ ft} + 35 \text{ ft} + 25 \text{ ft}$
$= 85 \text{ ft}$
Ken needs 85 feet of fencing. Since he has 115 feet, he has enough.
$115 - 85 = 30$
Ken has 30 feet of fencing left over.

33. $8^2 - 4 \cdot 2^3 = 64 - 4 \cdot 8$
$= 64 - 32$
$= 32$

35. $4^3 - 2 \cdot (15 - 12) = 4^3 - 2 \cdot 3$
$= 64 - 2 \cdot 3$
$= 64 - 6$
$= 58$

37. $2^4 \div 2^2 \cdot 8 = 16 \div 4 \cdot 8$
$= 4 \cdot 8$
$= 32$

39. $15.8 - 4 \cdot (0.6)^2$
$= 15.8 - 4 \cdot 0.36$
$= 15.8 - 1.44$
$= 14.36$

41. $\frac{5}{8}\left(\frac{1}{2}\right)^2 \cdot \frac{1}{3} = \frac{5}{8} + \frac{1}{4} \cdot \frac{1}{3}$
$= \frac{5}{8} + \frac{1}{12}$
$= \frac{5}{8} \cdot \frac{3}{3} + \frac{1}{12} \cdot \frac{2}{2}$
$= \frac{15}{24} + \frac{2}{24}$
$= \frac{17}{24}$

Problem Set 9.2

1. $A = 5 \text{ ft} \cdot 6 \text{ ft} = 30 \text{ ft}^2$

3. $A = 7 \text{ cm} \cdot 7 \text{ cm} = 49 \text{ cm}^2$

5. $A = 3.2 \text{ mm} \times 7.8 \text{ mm} = 24.96 \text{ mm}^2$

7. $a = 3\frac{1}{4} \text{ ft} \cdot 6\frac{1}{2} \text{ ft}$
$= \frac{13}{4} \text{ ft} \cdot \frac{13}{2} \text{ ft}$
$= \frac{169}{8} \text{ ft}^2 = 21\frac{1}{8} \text{ ft}^2$

9. $A = 30 \text{ cm} \cdot 70 \text{ cm} = 2100 \text{ cm}^2$

11. $A = 32 \text{ mm} \cdot 9 \text{ mm} = 288 \text{ mm}^2$

13.
$A = 6\frac{1}{4} \text{ m} \cdot 3\frac{1}{5} \text{ m} = \frac{25}{4} \text{ m} \cdot \frac{16}{5} \text{ m}$
$= \frac{400}{20} \text{ m}^2 = 20 \text{ m}^2$

15. $A = 74.35 \text{ yd} \times 100 \text{ yd}$
$= 7435 \text{ yd}^2$

17. $A = (24 \text{ ft})^2 = 576 \text{ ft}^2$

19. $A = (62.3 \text{ km})^2 = 3881.29 \text{ km}^2$

21. $A = \left(7\frac{3}{4} \text{ mm}\right)^2 = \left(\frac{31}{4} \text{ mm}\right)^2$
$= \frac{961}{16} \text{ mm}^2 = 60\frac{1}{16} \text{ mm}^2$

23. $A = 6 \text{ ft} \cdot 4 \text{ ft} = 24 \text{ ft}^2$
The area of the mirror is 24 square feet.

25. $A = 40 \text{ ft} \cdot 55 \text{ ft} = 2200 \text{ ft}^2$
Cost $= 2200 \cdot 56 = 123{,}200$
The cost of building the house is $123,200.

27. Dimensions of large rectange are 8 meters by 11 meters.
Dimensions of small rectange are $[11 - (7 + 2)] = 2$ meters by $(8 - 3) = 5$ meters
Area $= 8 \text{ m} \times 11 \text{ m} - 2 \text{ m} \times 5 \text{ m}$
$= 88 \text{ m}^2 - 10 \text{ m}^2$
$= 78 \text{ m}^2$

29. Area $= 42 \text{ ft} \cdot 56 \text{ ft} = 2352 \text{ ft}^2$
Cost $= 2352 \cdot 0.98 = 2304.96$
The area of the room is 2352 square feet and the cost of flooring is $2304.96.

Problem Set 9.2

31. Area of one face
$= 11 \text{ in.} \times 11 \text{ in.}$
$= 121 \text{ in.}^2$
Area of six faces
$= 6 \bullet 121 \text{ in}^2 = 726 \text{ in}^2$

33. $\frac{63 \not{days}}{1} \bullet \frac{1 \text{ week}}{7 \not{days}} = 9 \text{ weeks}$

35. $\frac{5 \not{miles}}{1} \bullet \frac{5280 \text{ feet}}{1 \not{mile}}$
$= 26{,}400 \text{ feet}$

37. $\frac{7 \not{lb}}{1} \bullet \frac{16 \text{ oz}}{1 \not{lb}} = 112 \text{ ounces}$

39. $\frac{1 \not{qt}}{1} \bullet \frac{2 \text{ pt}}{1 \not{qt}} = 2 \text{ pints}$

41. $\frac{324 \not{in.}}{1} \bullet \frac{1 \not{ft}}{12 \not{in.}} \bullet \frac{1 \text{ yd}}{3 \not{ft}}$
$= 9 \text{ yards}$

Problem Set 9.3

1. $A = 7 \text{ m} \bullet 4 \text{ m} = 28 \text{ m}^2$

3. $A = 2 \text{ ft} \bullet 9 \text{ ft} = 18 \text{ ft}^2$

5. $A = \frac{1}{2} \bullet 11 \text{ ft} \bullet 6 \text{ ft}$
$= \frac{66}{2} \text{ ft}^2 = 33 \text{ ft}^2$

7. $a = \frac{1}{2} \bullet 13 \text{ yd} \bullet 13 \text{ yd}$
$= \frac{169}{2} \text{ yd}^2 = 84\frac{1}{2} \text{ yd}^2$

9. $A = \frac{1}{2} \bullet 8 \text{ km} \bullet (12 \text{ km} + 16 \text{ km})$
$= \frac{1}{2} \bullet 8 \text{ km} \bullet 28 \text{ km}$
$= \frac{224 \text{ km}^2}{2} = 112 \text{ km}^2$

11. $A = \frac{1}{2} \bullet 10 \text{ yd}$
$\bullet (15 \text{ yd} + 25 \text{ yd})$
$= \frac{1}{2} \bullet 10 \text{ yd} \bullet 40 \text{ yd}$
$= \frac{400}{2} \text{ yd}^2 = 200 \text{ yd}^2$

13. $A = \frac{1}{2} \times 9.5 \text{ m} \times 7.3 \text{ m}$
$= 34.675 \text{ m}^2$

15. $A = \frac{1}{2} \times 16.1 \text{ yd}$
$\times (14.4 \text{ yd} + 25.2 \text{ yd})$
$= \frac{1}{2} \times 16.1 \text{ yd} \times 39.6 \text{ yd}$
$= 318.78 \text{ yd}^2$

17. $A = \frac{1}{2} \bullet 9 \text{ dm} \bullet (11 \text{ dm} + 14 \text{ dm})$
$= \frac{1}{2} \bullet 9 \text{ dm} \bullet 25 \text{ dm}$
$= \frac{225}{2} \text{ dm}^2 = 112\frac{1}{2} \text{ dm}^2$

19. $A = \frac{1}{2} \bullet 4 \text{ in.}$
$\bullet \left(6\frac{3}{4} \text{ in.} + 9\frac{1}{2} \text{ in.}\right)$
$= \frac{1}{2} \bullet 4 \text{ in.} \bullet 16\frac{1}{4} \text{ in.}$
$= 32\frac{1}{2} \text{ in.}^2$

21. $A = \frac{1}{2} \times 22.9 \text{ m} \times 14 \text{ m}$
$= 160.3 \text{ m}^2$

23. $A = 5 \text{ yd} \times 14.3 \text{ yd}$
$= 71.5 \text{ yd}^2$

25. $A = \frac{1}{2} \bullet 82 \text{ cm} \bullet 60 \text{ cm}$
$= 2460 \text{ cm}^2$
The area of the poster is 2460 square centimeters.

27. $A = \frac{1}{2} \bullet 11 \text{ m} \bullet (12 \text{ m} + 15 \text{ m})$
$= \frac{1}{2} \bullet 11 \text{ m} \bullet 27 \text{ m}$
$= 148\frac{1}{2} \text{ m}^2$

29. opposite

31. $A = 324 \text{ ft} \bullet 967 \text{ ft}$
$= 313{,}308 \text{ ft}^2$

Problem Set 9.3

33. $A = 14\text{ m} \cdot 5\text{ m} + \frac{1}{2} \cdot 8\text{ m} \cdot (16\text{ m} + 14\text{ m}) + 16\text{ m} \cdot 4\text{ m}$
$= 14\text{ m} \cdot 5\text{ m} + \frac{1}{2} \cdot 8\text{ m} \cdot 30\text{ m} + 16\text{ m} \cdot 4\text{ m}$
$= 70\text{ m}^2 + 120\text{ m}^2 + 64\text{ m}^2$
$= 254\text{ m}^2$

35. Area of sides
$= 2 \cdot 15\text{ ft} \cdot 60\text{ ft}$
$= 1800\text{ ft}^2$
Area of ends
$= 2 \cdot \left(30\text{ ft} \cdot 15\text{ ft} + \frac{1}{2} \cdot 30\text{ ft} \cdot 6\text{ ft}\right)$
$= 1080\text{ ft}^2$
Total area
$= 1800\text{ ft}^2 + 1080\text{ ft}^2$
$= 2880\text{ ft}^2$
The additional information needed is how many square feet 1 gallon of paint will cover.

37.
```
  4 hours 50 minutes
+19 hours 30 minutes
 23 hours 80 minutes
```
= 23 h + 1 h 20 min
= 24 h + 20 min
= 1 day 20 min

39.
```
 6          24
 7 pounds   8 ounces
-3 pounds  15 ounces
 3 pounds   9 ounces
```

41.
```
 4        5980
 5 miles  700 feet
-2 miles  850 feet
 2 miles 5130 feet
```

43.
```
  4 yd  6 ft  5 in.
×                 7
 28 yd 42 ft 35 in.
```
= 28 yd + 42 ft + 2 ft 11 in.
= 28 yd + 44 ft + 11 in.
= 28 yd + 14 yd 2 ft + 11 in
= 42 yd 2 ft 11 in.

45.
```
    3 miles   811 feet
 7)22 miles   397 feet
   21 miles
    1 mile = 5280 feet
             5677 feet
             5677 feet
                0
```

Problem Set 9.4

1. $d = 2 \cdot 9\text{ ft} = 18\text{ ft}$

3. $d = 2 \times 7.4\text{ cm} = 14.8\text{ cm}$

5. $r = \frac{42\text{ km}}{2} = 21\text{ km}$

7. $r = \frac{\frac{1}{2}\text{ in.}}{2} = \frac{1}{4}\text{ in.}$

9. $C \approx 2 \times 3.14 \times 9\text{ ft}$
$\approx 56.52\text{ ft}$

11. $C \approx 2 \times 3.14 \times 7.4\text{ cm}$
$\approx 46.472\text{ cm}$

13. $C \approx \frac{22}{7} \cdot 42\text{ km} \approx 132\text{ km}$

15. $C \approx \frac{22}{7} \cdot \frac{1}{2}\text{ in.} \approx \frac{11}{7}\text{ in.}$
$\approx 1\frac{4}{7}\text{ in.}$

17. $A \approx 3.14 \times (9\text{ ft})^2 \approx 254.34\text{ ft}^2$

19. $A \approx 3.14 \times (7.4\text{ cm})^2$
$\approx 171.9464\text{ cm}^2$

21. $A \approx \frac{22}{7} \cdot (21\text{ km})^2 \approx 1386\text{ km}^2$

23. $A \approx \frac{22}{7} \cdot \left(\frac{1}{4}\text{ in.}\right)^2 \approx \frac{22}{112}\text{ in.}^2$
$\approx \frac{11}{56}\text{ in.}^2$

25. $C \approx 3.14 \times 5\text{ in} \approx 15.7\text{ in.}$
$A \approx 3.14 \times (2.5\text{ in.})^2$
$\approx 19.625\text{ in.}^2$
The circumference of the lid is 15.7 inches and the area of the lid is 19.625 square inches.

Problem Set 9.4

27. $C \approx 3.14 \times 1.2 \text{ m} \approx 3.768 \text{ m}$
The circumference of the tree is 3.768 meters.

29. $A \approx 3.14 \times (16 \text{ ft})^2$
$\approx 803.84 \text{ ft}^2$
The area of the lawn is 803.84 square feet.

31. $C \approx 3.14 \times 25 \text{ cm} \approx 78.5 \text{ cm}$
$A \approx 3.14 \times (12.5 \text{ cm})^2$
$\approx 490.625 \text{ cm}^2$
The circumference of the burner is 78.5 centimeters and the area of the burner is 490.625 square centimeters.

33. Area of square
$= (12 \text{ ft})^2 = 144 \text{ ft}^2$
Area of circle
$\approx 3.14 \times (6 \text{ ft})^2$
$\approx 113.04 \text{ ft}^2$
Area of shaded region
$= 144 \text{ ft}^2 - 113.04 \text{ ft}^2$
$= 30.96 \text{ ft}^2$
The area of the flower beds is 30.96 square feet.

35. Area of large circle
$\approx 3.14 \times (9 \text{ cm})^2$
$\approx 254.34 \text{ cm}^2$
Area of small circles
$\approx 2 \times 3.14 \times (4.5 \text{ cm})^2$
$\approx 127.17 \text{ cm}^2$
Area of shaded region
$= 254.34 \text{ cm}^2 - 127.17 \text{ cm}^2$
$= 127.17 \text{ cm}^2$

37. twice, circumference

39. $A \approx 3.14 \times \left(\frac{96}{2} \text{ cm}\right)^2$
$\approx 7234.56 \text{ cm}^2$
The area of the trampoline is 7234.56 square centimeters.

41. $C \approx 3.14 \times 68 \text{ cm} \approx 213.52 \text{ cm}$
Revolutions
$= 9 \text{ km} \div 213.52 \text{ cm}$
$= 900{,}000 \text{ cm} \div 213.52 \text{ cm}$
≈ 4215.06
The tach moves 213.52 centimeters with each revolution. The tire would have to travel about 4215.06 revolutions for the tach to travel 9 kilometers.

43. 56 km = 56,000 m

45. 9.7 L = 9700 mL

47. 14.3 g = 14,300 mg

49. 54.8 m = 5480 cm

51. 825 mm = 82.5 cm

Problem Set 9.5

1. $V = 9 \text{ in.} \cdot 4 \text{ in.} \cdot 5 \text{ in.}$
$= 180 \text{ in.}^3$

3. $V = 11 \text{ ft} \cdot 11 \text{ ft} \cdot 11 \text{ ft}$
$= 1331 \text{ ft}^3$

5. $V \approx 3.14 \times (3 \text{ cm})^2 \times 15 \text{ cm}$
$\approx 423.9 \text{ cm}^3$

7. $V \approx 3.14 \times \left(\frac{3}{4} \text{ in.}\right)^2 \times 10 \text{ in.}$
$\approx 17.66 \text{ in.}^3$

9. $V \approx \frac{4}{3} \times 3.14 \times (10 \text{ yd})^3$
$\approx 4186.67 \text{ yd}^3$

11. $V \approx \frac{4}{3} \times 3.14 \times (5.4 \text{ cm})^3$
$\approx 659.25 \text{ cm}^3$

13. $V = 4 \text{ in.} \cdot 2 \text{ in.} \cdot 5 \text{ in.}$
$= 40 \text{ in.}^3$
The volume of the canister is 40 cubic inches.

15. $V \approx 3.14 \times (8 \text{ ft})^2 \times 30 \text{ ft}$
$\approx 6028.8 \text{ ft}^3$
The volume of the silo is 6028.8 cubic feet.

17. $V \approx \frac{4}{3} \times 3.14 \times (62.9 \text{ mm})^3$
$\approx 1{,}041{,}886.3 \text{ mm}^3$
The volume of the sphere is 1,041,886.3 cubic millimeters.

Problem Set 9.5

19. $V \approx \frac{4}{3} \times 3.14 \times \left(\frac{205}{2}\text{ ft}\right)^3$
$\approx 4{,}508{,}582.1\text{ ft}^3$
The volume of the spherical ride is 4,508,582.1 cubic feet.

21. Volume of right circular cylinder
$\approx 3.14 \times (3\text{ ft})^2 \times 15\text{ ft}$
$= 423.9\text{ ft}^3$
Volume of rectangular solid
$= 6\text{ ft} \bullet 6\text{ ft} \bullet 3\text{ ft}$
$= 108\text{ ft}^3$
Volume of support column
$= 423.9\text{ ft}^3 + 108\text{ ft}^3$
$= 531.0\text{ ft}^3$

23. Volume
$= 24\text{ in.} \bullet 12\text{ in.} \bullet 11\text{ in.}$
$= 3168\text{ in.}^3$
Weight
$= \frac{3168\text{ in.}^3}{1} \bullet \frac{0.0361\text{ lb}}{1\text{ in.}^3}$
$\approx 114.36\text{ lb}$
The weight of the water is about 114.36 pounds.

25. $18.02\text{ qt} \approx \frac{18.02\text{ qt}}{1} \bullet \frac{1\text{ L}}{1.06\text{ qt}} \approx 17\text{ L}$

27. $710\text{ mi} \approx \frac{710\text{ mi}}{1} \bullet \frac{1.61\text{ km}}{1\text{ mi}} \approx 1143.1\text{ km}$

29. $236.4\text{ in.} \approx \frac{236.4\text{ in.}}{1} \bullet \frac{1\text{ m}}{39.4\text{ in.}} \approx 6\text{ m}$

31. $F = \frac{9}{5} \bullet 52 + 32 = 125.6$

33. $C = \frac{5}{9} \bullet (77 - 32) = 25$

Chapter 9 Additional Exercises

1. $P = 8.2\text{ ft} + 12.4\text{ ft} + 7.1\text{ ft}$
$= 27.7\text{ ft}$

3. $P = 2 \bullet 7\text{ ft} + 2 \bullet 4\text{ ft} = 22\text{ ft}$

5. $P = 2 \times 37\text{ cm} + 2 \times 49.4\text{ cm}$
$= 172.8\text{ cm}$

7. $P = 4 \bullet 47\text{ mm} = 188\text{ mm}$

9. $P = 6\text{ m} + 7\text{ m} + (15\text{ m} - 10\text{ m}) + (8\text{ m} - 7\text{ m}) + 4\text{ m} + 8\text{ m} + 15\text{ m} + 14\text{ m} = 60\text{ m}$

11. $P = 2 \times 12\text{ ft} + 2 \times 15\frac{1}{4}\text{ ft}$
$= 54\frac{1}{2}\text{ ft}^2$
Cost $= 54\frac{1}{2} \bullet 0.98 = 53.41$
Aaron will spend $53.41 to put baseboards around the room.

13. $A = 48\text{ m} \bullet 72\text{ m} = 3456\text{ m}^2$

15. $A = 8\frac{1}{3}\text{ yd} \bullet 6\frac{2}{5}\text{ yd}$
$= \frac{25}{3}\text{ yd} \bullet \frac{32}{5}\text{ yd}$
$= 53\frac{1}{3}\text{ yd}^2$

17. $A = \left(4\frac{5}{7}\text{ in.}\right)^2 = \left(\frac{33}{7}\text{ in.}\right)^2$
$= \frac{1089}{49}\text{ in.}^2 = 22\frac{11}{49}\text{ in.}^2$

19. $A = 22.5\text{ in.} \times 39.5\text{ in.}$
$= 888.75\text{ in.}^2$
The area of the skylight is 888.75 square inches.

21. $A = 15\text{ m} \bullet 11\text{ m} = 165\text{ m}^2$

23. $A = 8\text{ mm} \bullet 24\text{ mm} = 192\text{ mm}^2$

25. $A = \frac{1}{2} \bullet 17\text{ km} \bullet 3\text{ km}$
$= 25\frac{1}{2}\text{ km}^2$

27. $A = \frac{1}{2} \bullet 9\text{ ft} \bullet (8\text{ ft} + 12\text{ ft})$
$= \frac{1}{2} \bullet 9\text{ ft} \bullet 20\text{ ft}$
$= 90\text{ ft}^2$

Chapter 9 Additional Exercises

29. $A = \frac{1}{2} \bullet 4\frac{3}{5}$ in. $\bullet \left(6\frac{1}{2} \text{ in.} + 8\frac{1}{4} \text{ in.}\right)$
$= \frac{1}{2} \bullet \frac{23}{5} \text{ in.} \bullet \frac{59}{4} \text{ in.}$
$= 33\frac{37}{40} \text{ in.}^2$

31. Area $= \frac{1}{3} \text{ ft} \bullet 16 \text{ ft} = 5\frac{1}{3} \text{ ft}^2$
Decking $= 256 \text{ ft}^2 + 5\frac{1}{3} \text{ ft}^2$
$= 48$
The amount of decking needed is 48 pieces.

33. $C \approx 2 \times 3.14 \times 12$ in.
≈ 75.36 in.

35. $C \approx 3.14 \times 16.3 \text{ mm} \approx 51.182 \text{ mm}$

37. $A \approx 3.14 \times (12 \text{ in.})^2$
$\approx 452.16 \text{ in.}^2$

39. $A \approx 3.14 \times \left(\frac{16.3}{2} \text{ mm}\right)^2$
$\approx 208.57 \text{ mm}^2$

41. $V = 32 \text{ cm} \bullet 34 \text{ cm} \bullet 10 \text{ cm}$
$= 10{,}880 \text{ cm}^3$

43. $V = 12 \text{ m} \bullet 11 \text{ m} \bullet 3 \text{ m} = 396 \text{ m}^3$

45. $V \approx 3.14 \times (8.4 \text{ mm})^2 \times 6.3 \text{ mm}$
$\approx 1395.82 \text{ mm}^3$

47. $V \approx \frac{4}{3} \times 3.14 \times (8 \text{ ft})^3$
$\approx 2143.57 \text{ ft}^3$

49. $V \approx \frac{4}{3} \times 3.14 \times (30 \text{ in.})^3$
$\approx 113{,}040 \text{ in}^3$

Chapter 9 Practice Test

1. $P = 2 \times 9.4 \text{ ft} + 2 \times 12 \text{ ft}$
$= 42.8 \text{ ft}^2$

3. $A = 14.8 \text{ m} \times 7 \text{ m} = 103.6 \text{ m}^2$

5. $A = \frac{1}{2} \bullet 5 \text{ ft} \bullet (7 \text{ ft} + 9 \text{ ft})$
$= \frac{1}{2} \bullet 5 \text{ ft} \bullet 16 \text{ ft} = 40 \text{ ft}^2$

7. $A \approx 3.14 \times \left(\frac{28}{2} \text{ m}\right)^2$
$\approx 615.44 \text{ m}^2$

9. $V \approx 3.14 \times \left(\frac{18.4}{2} \text{ cm}\right)^2 \times 9 \text{ cm}$
$\approx 2391.9264 \text{ cm}^3$

11. $A = 3 \text{ ft} \times 4.5 \text{ ft} = 13.5 \text{ ft}^2$
The window has an area of 13.5 square feet.

13. $A \approx 3.14 \times (15 \text{ ft})^2 \approx 706.5 \text{ ft}^2$
Lime $= \frac{40 \text{ lb}}{1000 \text{ ft}^2} \bullet \frac{706.5 \text{ ft}^2}{1}$
$= 28.26 \text{ lb} \approx 28.3 \text{ lb}$
28.3 pounds of lime are needed.

15. $V \approx \frac{4}{3} \times 3.14 \times \left(\frac{24}{2} \text{ cm}\right)^3$
$\approx 7234.56 \text{ cm}^3$
The volume of the balloon is 7234.56 cubic centimeters.

Problem Set 10.1

1. A depth of 48 feet below sea level may be represented by -48 feet.

3. A height of 11,324 feet above sea level may be represented by 11,324 feet.

5. The loss may be represented by -$74.

7. The opposite of 9 is -9

9. The opposite of -2 is 2

11. The opposite of $-\frac{3}{5}$ is $\frac{3}{5}$

13. The opposite of 5.6 is -5.6

15. The opposite of $\frac{11}{3}$ is $-\frac{11}{3}$

17. The opposite of -14.8 is 14.8.

19. $-(-4) = 4$

21. $-(3) = -3$

Problem Set 10.1

23. $-(-18) = 18$

25. $-\left(\frac{5}{6}\right) = -\frac{5}{6}$

27. $-(-4.8) = 4.8$

29. $-(18.1) = -18.1$

31. $8 > 5$

33. $-5 < -3$

35. $-4 < 5$

37. $-1 > -3$

39. $0 > -5$

41. $-4 < -2$

43. $|8| = 8$

45. $|-7| = 7$

47. $|0| = 0$

49. $|-48| = 48$

51. $-|-4| = -4$

53. $\left|\frac{3}{4}\right| = \frac{3}{4}$

55. $|-8.7| = 8.7$

57. $-|-16.3| = -16.3$

59. zero

61. -10

63. $-|7|, -3.2, -\frac{5}{8}, \frac{11}{4}, |-6|, 8$

65. $-\frac{11}{7} > -\left|\frac{12}{5}\right|$, since $-\frac{11}{7} > -\frac{12}{5}$

67. $P = 6 \text{ in.} + 10 \text{ in.} + 6 \text{ in.} + 10 \text{ in.} = 32 \text{ in.}$

69. $P = 2 \bullet 3\frac{7}{8} \text{ m} + 2 \bullet 6 \text{ m} = 19\frac{3}{4} \text{ m}$

71. $P = 2 \times 43.8 \text{ cm} + 2 \times 67.18 \text{ cm} = 221.8 \text{ cm}$

73. $P = 4 \bullet 6\frac{3}{4} \text{ mm} = 27 \text{ mm}$

75. $P = 4 \times 366.7 \text{ ft} = 1466.8 \text{ ft}$

Problem Set 10.2

1. $(+7) + (+2) = +(7+2) = +9 = 9$

3. $(-3) + (-4) = -(3+4) = -7$

5. $(-9) + (-7) = -(9+7) = -16$

7. $(+7) + (+6) = +(7+6) = +13 = 13$

9. $(-3.4) + (-7.5) = -(3.4+7.5) = -10.9$

11. $\left(-\frac{1}{3}\right) + \left(-\frac{1}{4}\right) = -\left(\frac{1}{3} + \frac{1}{4}\right) = -\frac{7}{12}$

13. $(+5) + (-3) = +(5-3) = +2 = 2$

15. $(-8) + (+3) = -(8-3) = -5$

17. $(-10) + (+6) = -(10-6) = -4$

19. $(+17) + (-8) = +(17-8) = +9 = 9$

21. $(-8.9) + (+6.3) = -(8.9-6.3) = -2.6$

23. $\left(-\frac{7}{8}\right) + \left(+\frac{5}{4}\right) = +\left(\frac{5}{4} - \frac{7}{8}\right) = +\frac{3}{8} = \frac{3}{8}$

25. $-3 + 3 = 0$

27. $5 + 0 = 5$

29. $(-8) + 0 = -8$

31. $-6 + 0 = -6$

33. $8 + 4 = (+8) + (+4) = +(8 + 4) = +12 = 12$

35. $-9 + 3 = (-9) + (+3) = -(9 - 3) = -6$

37. $6 + (-8) = (+6) + (-8) = -(8 - 6) = -2$

39. $-7 + (-3) = (-7) + (-3) = -(7 + 3) = -10$

41. $-18 + 18 = 0$

Problem Set 10.2

43. $54 + (-54) = 0$

45. $-15.6 + 15.6 = 0$

47. $18 + (-5) = (+18) + (-5)$
$= +(18 - 5) = +13 = 13$

49. $(-12) + (-23) = -(12 + 23)$
$= -35$

51. $-46 + 20 = (-46) + (+20)$
$= -(46 - 20) = -26$

53. $-\frac{3}{8} + \frac{3}{8} = 0$

55. $\frac{5}{6} + \frac{1}{3} = \frac{5}{6} + \frac{1}{3} \bullet \frac{2}{2} = \frac{5}{6} + \frac{2}{6}$
$= \frac{7}{6} = 1\frac{1}{6}$

57. $-74 + (-22) = (-74) + (-22)$
$= -(74 + 22) = -96$

59. $-9 + 23 = (-9) + (+23)$
$= +(23 - 9) = +14 = 14$

61. $-3.8 + (-4.6) + 9.3$
$= -(3.8 + 4.6) + 9.3$
$= -8.4 + 9.3 = 0.9$

63. $-22 + 22 = 0$

65. $10 + (-18) = (+10) + (-18)$
$= -(18 - 10) = -8$

67. $-33 + 4 = (-33) + (+4)$
$= -(33 - 4) = -29$

69. $-\frac{7}{8} + \left(-\frac{9}{16}\right) = \left(-\frac{14}{16}\right) + \left(-\frac{9}{16}\right)$
$= -\left(\frac{14}{16} + \frac{9}{16}\right) = -\frac{23}{16}$
$= -1\frac{7}{16}$

71. $74 + (-24) = (+74) + (-24)$
$= +(74 - 24) = +50$
$= 50$

73. $(-3) + 7 + (-1) + 6$
$= +(7 - 3) + (-1) + 6$
$= (+4) + (-1) + 6$
$= +(4 - 1) + 6$
$= 3 + 6 = 9$

75.
$$\begin{array}{r} -22 \\ -18 \\ -30 \\ -\ 7 \\ \underline{-12} \\ -89 \end{array}$$

77.
$$\begin{array}{r} -15 \\ 62 \\ 28 \\ -13 \\ \underline{-\ 8} \end{array} \quad \begin{array}{r} 62 \\ \underline{28} \\ 90 \end{array} \quad \begin{array}{r} -15 \\ -13 \\ \underline{-\ 8} \\ -36 \end{array} \quad \begin{array}{r} 90 \\ \underline{-36} \\ 54 \end{array}$$
The sum is 54.

79. $(+48) + (-76) = -(76 - 48)$
$= -28$
Juma's account is overdrawn by \$28.

81. add

83. zero

85.
$$\begin{array}{r} -16{,}015 \\ -23{,}814 \\ 78{,}190 \\ -31{,}849 \\ \underline{-51{,}782} \\ 58{,}294 \end{array}$$

87.
$$\begin{array}{r} -\frac{5}{8} \\ \frac{3}{4} \\ \underline{-\frac{7}{12}} \end{array} \quad \begin{array}{r} -\frac{15}{24} \\ \frac{18}{24} \\ \underline{-\frac{14}{24}} \\ -\frac{11}{24} \end{array}$$

89. $A = 6.9 \text{ in.} \times 17.4 \text{ in.}$
$= 120.06 \text{ in.}^2$

91. $A = \left(6\frac{3}{4} \text{ yd}\right)^2 = \left(\frac{27}{4} \text{ yd}\right)^2$
$= \frac{729}{16} \text{ yd}^2 = 45\frac{9}{16} \text{ yds}^2$

93. $A = 3\frac{1}{8} \text{ ft} \bullet 6\frac{1}{4} \text{ ft}$
$= \frac{25}{8} \text{ ft} \bullet \frac{25}{4} \text{ ft}$
$= \frac{625}{32} \text{ ft}^2 = 19\frac{17}{32} \text{ ft}^2$

95. $A = (9.5 \text{ m})^2 = 90.25 \text{ m}^2$

Problem Set 10.2

97. $A = \left(3\frac{1}{4}\text{ ft}\right)^2 = \left(\frac{13}{4}\text{ ft}\right)^2$
$= \frac{169}{16}\text{ ft}^2 = 10\frac{9}{16}\text{ ft}^2$

Problem Set 10.3

1. $(+5) - (+3) = (+5) + (-3) = 2$

3. $(-6) - (+3) = (-6) + (-3) = -9$

5. $(+4) - (+9) = (+4) + (-9) = -5$

7. $(-6.8) - (+4.2)$
$= (-6.8) + (-4.2) = -11$

9. $(-8) - (-1) = (-8) + (+1) = -7$

11. $(-6) - (+6) = (-6) + (-6) + -12$

13. $(-5) - (-5) = (-5) + (+5) = 0$

15. $(+8) - (-8) = (+8) + (+8) = 16$

17. $(+9) - (+3) = (+9) + (-3) = 6$

19. $\left(-\frac{9}{5}\right) - \left(-\frac{6}{5}\right) = \left(-\frac{9}{5}\right) + \left(+\frac{6}{5}\right) = -\frac{3}{5}$

21. $(+4) - (-1) = (+4) + (+1) = 5$

23. $(-16) - (+10) = (-16) + (-10)$
$= -26$

25. $0 - 7 = 0 + (-7) = -7$

27. $3 - (-3) = (+3) + (+3) = 6$

29. $9 - 4 = (+9) + (-4) = 5$

31. $-8 - 6 = (-8) + (-6) = -14$

33. $8 - (-6) = (+8) + (+6) = 14$

35. $-4 - (-2) = (-4) + (+2) = -2$

37. $-8 - 0 = (-8) + (0) + -8$

39. $0 - (-7) = 0 + (+7) = 7$

41. $-7 - (-6) = (-7) + (+6) = -1$

43. $\frac{3}{5} - \frac{4}{3} - \left(+\frac{3}{5}\right) + \left(-\frac{4}{3}\right)$
$= \left(+\frac{9}{15}\right) + \left(-\frac{20}{15}\right) = -\frac{11}{15}$

45. $-\frac{9}{4} - \frac{3}{8} = \left(-\frac{9}{4}\right) + \left(-\frac{3}{8}\right)$
$= \left(-\frac{18}{8}\right) + \left(-\frac{3}{8}\right) = -\frac{21}{8}$
$= -2\frac{5}{8}$

47. $4 - 20 = (+4) + (-20) = -16$

49. $-16 - 62 = (-16) + (-62) = -78$

51. $-25.2 - 34 = (-25.2) + (-34)$
$= -59.2$

53. $-3 - 28 = (-3) + (-28) = -31$

55. $17 - 54 = (+17) + (-54) = -37$

57. $60 - (-60) = (+60) + (+60) = 120$

59. $19 - 33 = (+19) + (-33) = -14$

61. $\frac{5}{16} - \frac{7}{8} = \left(+\frac{5}{16}\right) + \left(-\frac{7}{8}\right)$
$= \left(+\frac{5}{16}\right) + \left(-\frac{14}{16}\right) = -\frac{9}{16}$

63. $17.8 - (-16.1)$
$= (+17.8) + (+16.1)$
$= 33.9$

65. $74 - (-74) = (+74) + (+74) = 148$

67. $0 - 215 = 0 + (-215) = -215$

69. $19 - (-3) + 7 + 8 - 6$
$= 19 + (+3) + 7 + 8 + (-6)$
$= 31$

71. $-36 - 14 + 22 - (-15)$
$= -36 + (-14) + 22 + (+15)$
$= -13$

73. $-42 - (-28) - (-34) - 52$
$= -42 + (+28) + (+34) + (-52)$
$= -32$

75. opposite

77. $624 + (-320) + (-61) + 357$
$+ 28 + (-108) = 520$
Karen's account had \$520.

79. $-84.1 - (-32.8) - 46.85$
$- (-89.3) - 68.79$
$= -84.1 + (+32.8) + (-46.85)$
$+ (+89.3) + (-68.79)$
$= -77.64$

81. true

83. $\frac{3}{10} \div \frac{7}{20} = \frac{3}{10} \bullet \frac{20}{7} = \frac{3}{\cancel{10}} \bullet \frac{2 \bullet \cancel{10}}{7} = \frac{6}{7}$

85. $\frac{7}{15} \div \frac{14}{1} = \frac{7}{15} \bullet \frac{1}{14} = \frac{\cancel{7}}{15} \bullet \frac{1}{2 \bullet \cancel{7}} = \frac{1}{30}$

87. $\frac{3}{4} \div \frac{9}{16} = \frac{3}{4} \bullet \frac{16}{9} = \frac{\cancel{3}}{\cancel{4}} \bullet \frac{\cancel{4} \bullet 4}{\cancel{3} \bullet 3} = \frac{4}{3} = 1\frac{1}{3}$

89. $A = \frac{1}{2} \times 18.4 \text{ cm} \times 9.2 \text{ cm} = 84.64 \text{ cm}^2$

Problem Set 10.4

1. $-7 \bullet 4 = -28$

3. $-8 \bullet 2 = -16$

5. $6 \bullet (-3) = -18$

7. $5 \bullet (-3) = -15$

9. $-12 \div 4 = -3$

11. $-9 \div 3 = -3$

13. $35 \div (-7) = -5$

15. $36 \div (-6) = -6$

17. $-5.7 \times 2.3 = -13.11$

19. $\frac{3}{5} \bullet \left(-\frac{5}{9}\right) = -\frac{1}{3}$

21. $-6 \bullet (-7) = 42$

23. $-3 \bullet (-10) = 30$

25. $-9 \bullet (-5) = 45$

27. $-4 \bullet (-11) = 44$

29. $-40 \div (-8) = 5$

31. $\frac{-24}{-2} = 12$

33. $-16.2 \div (-2) = 8.1$

35. $-\frac{1}{5} \div \left(-\frac{8}{5}\right) = -\frac{1}{5} \bullet \left(-\frac{5}{8}\right) = \frac{1}{8}$

37. $\left(-\frac{7}{8}\right) \div \left(-\frac{7}{24}\right) = \left(-\frac{7}{8}\right) \bullet \left(-\frac{24}{7}\right) = 3$

39. $7 \bullet (-12) = -84$

41. $-9 \bullet (-11) = 99$

43. $-8.4 \times (-5.2) = 43.68$

45. $-\frac{7}{8} \bullet \left(-\frac{16}{5}\right) = \frac{14}{5} = 2\frac{4}{5}$

47. $-35 \bullet 50 = -1750$

49. $34 \bullet (-2) = -68$

51. $\frac{-33}{3} = -11$

53. $-8.5 \times 4 = -34$

55. $-\frac{5}{11} \bullet \left(-\frac{22}{15}\right) = \frac{2}{3}$

57. $-200 \div 25 = -8$

59. $-\frac{3}{5} \div \left(-\frac{7}{10}\right) = -\frac{3}{5} \bullet \left(-\frac{10}{7}\right) = \frac{6}{7}$

61. $10.26 \div (-3.8) = -2.7$

63. $\frac{-12}{11} = -1\frac{1}{11}$

65. $\frac{-15}{-10} = \frac{3}{2} = 1\frac{1}{2}$

67. $4 \bullet (-7) \bullet (-3) = -28 \bullet (-3) = 84$

69. $-3 \bullet (-8) \bullet (-6) = 24 \bullet (-6) = -144$

71. $-2 \bullet (-3) \bullet (-4) \bullet (-1) = 6 \bullet (-4) \bullet (-1) = -24 \bullet (-1) = 24$

73. $\frac{-2}{3} = \frac{2}{-3} = -\frac{2}{3}$

75. $\frac{9}{-5} = \frac{-9}{5} = -\frac{9}{5}$

Problem Set 10.4

77. $-\frac{7}{11} = \frac{-7}{11} = \frac{7}{-11}$

79. $\frac{-10}{7} = \frac{10}{-7} = -\frac{10}{7}$

81. $\frac{16}{-3} = \frac{-16}{3} = -\frac{16}{3}$

83. $-\frac{4}{7} = \frac{-4}{7} = \frac{4}{-7}$

85. $3 \bullet (-46) = -138$
John's total debt was $138.

87. two

89. $456.68 \bullet (-46.6)$
$= -21{,}281.288$

91. true

93. $C \approx 3.14 \times 15 \text{ cm} \approx 47.1 \text{ cm}$

95. $C \approx 2 \times 3.14 \times 4.2 \text{ in}$
$\approx 26.376 \text{ in}$

97. $r = \frac{15 \text{ cm}}{2} = 7.5 \text{ cm}$

99. $A \approx 3.14 \times (7.5 \text{ cm})^2$
$\approx 176.625 \text{ cm}^2$

101. $A \approx 3.14 \times (4.2 \text{ in})^2$
$\approx 55.3896 \text{ in}^2$

Problems Set 10.4A

1. $-7 - 4 = -7 + (-4) = -11$

3. $-5 \bullet (-6) = 30$

5. $9 + (-6) = 5$

7. $7 \bullet (-5) = -35$

9. $-5 + (-8) = -13$

11. $-16 \div (-8) = 2$

13. $8 - (-3) = 8 + (+3) = 11$

15. $-36 \div 6 = -6$

17. $9 - 15 = 9 + (-15) = -6$

19. $6 \bullet (-8) = -48$

21. $16 + (-28) = -12$

23. $54 \div (-9) = -6$

25. $(-7) \bullet (-3) = 21$

27. $-12 - (-4) = -12 + (+4) = -8$

29. $\frac{-48}{-3} = 16$

31. $\frac{-11}{-7} = \frac{11}{7} = 1\frac{4}{7}$

33. $-14.5 - 7.2 = -14.5 + (-7.2)$
$= -21.7$

35. $-2.1 \times (-6.3) = 13.23$

37. $\frac{3}{8} - \left(-\frac{11}{4}\right) = \frac{3}{8} + \left(\frac{22}{8}\right) = \frac{25}{8} = 3\frac{1}{8}$

39. $\frac{5}{6} \bullet \left(-\frac{12}{25}\right) = -\frac{2}{5}$

41. $56 - (-15) = 56 + (+15) = 71$
The difference is 71° F.

43. $-15 - (-24) = -15 + (+24)$
$= 9$
The difference is 9° F.

Problem Set 10.5

1. a. $5 \bullet (2 + 3) = 5 \bullet 5$
$= 25$
b. $5 \bullet 2 + 5 \bullet 3$
$= 10 + 15 = 25$

3. a. $7 \bullet (5+1) = 7 \bullet 6 = 42$
b. $7 \bullet 5+7 \bullet 1 = 35+7 = 42$

5. $3(x+4) = 3 \bullet x+3 \bullet 4 = 3x+12$

7. $8(x-1) = 8 \bullet x-8 \bullet 1 = 8x-8$

9. $-2(x-3) = -2 \bullet x+(-2) \bullet (-3)$
$= -2x+6$

11. $-2(3x+y) = -2 \bullet 3x+(-2) \bullet y$
$= -6x-2y$

13. $4(2x-y+7)$
$= 4 \bullet 2x-4 \bullet y+4 \bullet 7$
$= 8x-4y+28$

15. x, 2

Problem Set 10.5

17. $3x$, $4y$

19. x

21. $7x$, $-2y$, 5

23. $7x + 2x = (7 + 2)x$

25. $8x - 4x = (8 - 4)x$

27. $3x + x = (3 + 1)x$

29. $1.5y + 3.7y = (1.5 + 3.7)y$

31. $5x + 3x - 7x = (5 + 3 - 7)x$

33. $8y + 2y = (8 + 2)y = 10y$

35. $15a - 8a = (15 - 8)a = 7a$

37. $5x + x = (5 + 1)x = 6x$

39. $x - 14x = (1 - 14)x = -13x$

41. $5x+2y-3x+7y = 5x-3x+2y+7y$
$= (5-3)x+(2+7)y = 2x+9y$

43. $x-3.1x = (1-3.1)x = -2.1x$

45. $3.4x-7.2y-x+5.9y$
$= 3.4x-x+5.9y-7.2y$
$= (3.4-1)x+(5.9-7.2)y$
$= 2.4x-1.3y$

47. $\frac{5}{8}x+\frac{1}{8}y+\frac{7}{8}x-\frac{3}{8}y$
$= \frac{5}{8}x+\frac{7}{8}x+\frac{1}{8}y-\frac{3}{8}y$
$= \left(\frac{5}{8}+\frac{7}{8}\right)x+\left(\frac{1}{8}-\frac{3}{8}\right)y$
$= \frac{12}{8}x-\frac{2}{8}y$
$= \frac{3}{2}x-\frac{1}{4}y$

49. $3x+7 = 3(-2)+7 = -6+7 = 1$

51. $5x-6 = 5(-2)-6 = -10-6 = -16$

53. $4y+5-2y+3$
$= 4(2)+5-2(2)+3$
$= 8+5-4+3 = 12$

55. $4y+5-2y+3$
$= 4(-4)+5-2(-4)+3$
$= -16+5+8+3 = 0$

57. $d = 55(7) = 385$
The car travels 385 miles.

59. variable

61. like

63. a. $9.3 \times (7.8+16.5)$
$= 9.3 \times 24.3$
$= 225.99$

b. $9.3 \times 7.8+9.3 \times 16.5$
$= 72.54+153.45$
$= 225.99$

65. $3x+36 = 3 \bullet x+3 \bullet 12 = (x+12)3$

67. $bx + by - bz = (x + y - z)b$

69. $V \approx 3.14 \times \left(\frac{14}{2}\text{ ft}\right)^2 \times 9\text{ ft}$
$\approx 1384.74\text{ ft}^3$

71. $V = 9.3\text{ in} \times 7\text{ in} \times 5.5\text{ in}$
$= 358.05\text{ in}^3$
The volume of the box is 358.05 cubic inches.

73. $V \approx \frac{4}{3} \times 3.14 \times (3\text{ in.})^3$
$\approx 113.04\text{ in.}^3$

75. $V \approx \frac{4}{3} \times 3.14 \times (9\text{ cm})^3$
$= 3052.08\text{ cm}^3$
The sphere has a volume of 3052.08 cubic centimeters.

77. $V \approx 3.14 \times \left(\frac{20}{2}\text{ ft}\right)^2 \times 8\text{ ft}$
$= 2512\text{ ft}^3$
The right circular cylinder has a volume of 2512 cubic feet.

Chapter 10 Additional Exercises

1. The opposite of 7.8 is -7.8

3. The opposite of $-\frac{4}{5}$ is $\frac{4}{5}$

5. $-(6.2) = -6.2$

7. $-(-21.8) = 21.8$

9. $-\{-[-(-4.2)]\} = 4.2$

Chapter 10 Additional Exercises

11. $-6 < 3$

13. $-5.7 > -8.2$

15. $|-4| = 4$

17. $\left|-\frac{8}{5}\right| = \frac{8}{5}$

19. $16+(-9) = +(16-9) = 7$

21. $-14 + 14 = 0$

23. $-12 + 7 = -(12 - 7) = -5$

25. $-8 + 0 = -8$

27. $-4 + 9 + (-3) + (-8) = -6$

29. $12 - 9 = 12 + (-9) = 3$

31. $-8 - 6 = -8 + (-6) = -14$

33. $-6 - (-5) = -6 + (+5) = -1$

35. $\frac{5}{12} - \frac{7}{8} = \frac{5}{12} + \left(-\frac{7}{8}\right) = \frac{10}{24} + \left(-\frac{21}{24}\right)$
$= -\frac{11}{24}$

37. $23-(-4)+6-8+2$
$= 23+(+4)+6+(-8)+2$
$= 27$

39. $-19-(-74)-(-82)-95-(-16)$
$= -19+(+74)+(+82)$
$+(-95)+(+16)$
$= 58$

41. $(-7) \bullet (-6) = 42$

43. $\frac{-72}{8} = -9$

45. $-\frac{8}{7} \bullet \left(-\frac{21}{16}\right) = \frac{3}{2} = 1\frac{1}{2}$

47. $-7.5 \times 6.4 = -48$

49. $7(x+4) = 7 \bullet x+7 \bullet 4 = 7x+28$

51. $-2(y+5) = -2 \bullet y+(-2) \bullet 5$
$= -2y-10$

53. y

55. $7a,\ -4b$

57. $9x+6x = (9+6)x$

59. $3.1x+4.5x = (3.1+4.5)x$

61. $9x-x = (9-1)x = 8x$

63. $5.4y-8.2y = (5.4-8.2)y$
$= -2.8y$

65. $\frac{3}{5}x-\frac{7}{10}y-\frac{9}{4}x-\frac{8}{15}y$
$= \frac{3}{5}x-\frac{9}{4}x-\frac{7}{10}y-\frac{8}{15}y$
$= \left(\frac{3}{5}-\frac{9}{4}\right)x+\left(-\frac{7}{10}-\frac{8}{15}\right)y$
$= \left(\frac{12}{20}-\frac{45}{20}\right)x+\left(-\frac{21}{30}-\frac{16}{30}\right)y$
$= -\frac{33}{20}x-\frac{37}{30}y$

67. $9x-4-3x+2$
$= 9(-3)-4-3(-3)+2$
$= -27-4+9+2$
$= -20$

Chapter 10 Practice Test

1. $-(-8) = 8$

3. $|-7| = 7$

5. $5.7 + (-3.8) = 1.9$

7. $2 - (-5) = 2 + (+5) = 7$

9. $\frac{3}{8} - \frac{5}{6} = \frac{9}{24} + \left(-\frac{20}{24}\right) = -\frac{11}{24}$

11. $-72 \div (-8) = 9$

13. $-6.8 \times 7.2 = -48.96$

15. $-2(y - 8) = -2y + 16$

17. $7x+6-9x+14$
$= 7(-1)+6-9(-1)+14$
$= -7+6+9+14 = 22$

Chapters 9 and 10 Cumulative Review

1. $P = 2 \times 16.7 \text{ ft}+2 \times 11.3 \text{ ft}$
$= 56 \text{ ft}$

3. $A = (16 \text{ m})^2 = 256 \text{ m}^2$

Chapters 9 and 10 Cumulative Review

5. $A = \frac{1}{2} \times 9.7 \text{ mm} \times 4 \text{ mm}$
$= 19.4 \text{ mm}^2$

7. $C \approx 3.14 \times 18 \text{ ft} \approx 56.52 \text{ ft}$

9. $V = 6 \text{ m} \bullet 6 \text{ m} \bullet 6 \text{ m} = 216 \text{ m}^3$

11. $V \approx \frac{4}{3} \times 3.14 \times (6.1 \text{ in.})^3$
$\approx 950.29 \text{ in.}^3$

13. $C \approx 3.14 \times 6 \text{ m} \approx 18.84 \text{ m}$
18.84 meters of binding is needed.

15. $V \approx \frac{4}{3} \times 3.14 \times \left(\frac{6}{2} \text{ in.}\right)^3$
$\approx 113.04 \text{ in.}^3$

17. $0 > -4$

19. $|-16.3| = 16.3$

21. $-9.8 + (-4.3) = -14.1$

23. $-14 - 3 = -14 + (-3) = -17$

25. $\frac{7}{3} - \left(-\frac{5}{8}\right) = \frac{7}{3} + \left(+\frac{5}{8}\right)$
$= \frac{56}{24} + \frac{15}{24} = \frac{71}{24} = 2\frac{23}{24}$

27. $\frac{-81}{9} = -9$

29. $-4(x-9) = -4x+36$

31. $7a-8a-6b-4b-9$
$= (7-8)a+(-6-4)b-9$
$= -a-10b-9$

Problem Set 11.1

1. $x+3 = 2$
$x+3+(-3) = 2+(-3)$
$x = -1$

3. $8+y = -4$
$8+(-8)+y = -4+(-8)$
$y = -12$

5. $x-9 = 4$
$x-9+9 = 4+9$
$x = 13$

7. $x-3.5 = 8.2$
$x-3.5+3.5 = 8.2+3.5$
$x = 11.7$

9. $\frac{1}{2} = \frac{5}{2}+a$
$\frac{1}{2}+\left(-\frac{5}{2}\right) = \frac{5}{2}+\left(-\frac{5}{2}\right)+a$
$-\frac{4}{2} = a$
$-2 = a$

11. $x-\frac{3}{5} = \frac{2}{3}$
$x-\frac{3}{5}+\frac{3}{5} = \frac{2}{3}+\frac{3}{5}$
$x = \frac{10}{15}+\frac{9}{15}$
$x = \frac{19}{15}$
$x = 1\frac{4}{15}$

13. $p+9 = 20$
$p+9+(-9) = 20+(-9)$
$p = 11$

15. $-5.9+x = 6.8$
$-5.9+5.9+x = 6.8+5.9$
$x = 12.7$

17. $x+\frac{1}{5} = 1$
$x+\frac{1}{5}+\left(-\frac{1}{5}\right) = 1+\left(-\frac{1}{5}\right)$
$x = \frac{5}{5}-\frac{1}{5}$
$x = \frac{4}{5}$

19. $x+\frac{3}{5} = -\frac{4}{15}$
$x+\frac{3}{5}+\left(-\frac{3}{5}\right) = -\frac{4}{15}+\left(-\frac{3}{5}\right)$
$x = -\frac{4}{15}+\left(-\frac{9}{15}\right)$
$x = -\frac{13}{15}$

21. $-15 = p-18$
$-15+18 = p-18+18$
$3 = p$

Problem Set 11.1

23. $a - \frac{3}{7} = \frac{4}{21}$

$a - \frac{3}{7} + \frac{3}{7} = \frac{4}{21} + \frac{3}{7}$

$a = \frac{4}{21} + \frac{9}{21}$

$a = \frac{13}{21}$

25. $x + 113.4 = 9.8$

$x + 13.4 + (-13.4)$
$= 9.8 + (-13.4)$

$x = -3.6$

27. $x - 3\frac{1}{4} = 8$

$x - 3\frac{1}{4} + 3\frac{1}{4} = 8 + 3\frac{1}{4}$

$x + 1\frac{1}{4}$

29. $3\frac{1}{4} + a = 6\frac{1}{3}$

$3\frac{1}{4} + \left(-3\frac{1}{4}\right) + a = 6\frac{1}{3} + \left(-3\frac{1}{4}\right)$

$a = \frac{19}{3} + \left(-\frac{13}{4}\right)$

$a = \frac{76}{12} + \left(-\frac{39}{12}\right)$

$a = \frac{37}{12}$

$a = 3\frac{1}{12}$

31. $485 = C + 94$

$485 + (-94) = C + 94 + (-94)$

$391 = C$

The store paid \$391.

33. $15{,}800 = P + 1200$

$15{,}800 + (-1200)$
$= P + 1200 + (-1200)$

$14{,}600 = P$

The principal was \$14,600.

35. equal

37. same

39. $12{,}894 + y = -39{,}487$

$12{,}894 + (-12{,}894) + y$
$= -39{,}487 + (-12{,}894)$

$y = -52{,}381$

41. $5(x-2) = 5x - 15$

$5x - 10 = 5x - 15$

$5x + (-5x) - 10$
$= 5x + (-5x) - 15$

$-10 = -15$

This is not true. Therefore, there is no solution.

43. $-8.3 + (-9.7) = -18$

45. $-9 + (-4) + 25 + (-8) = 4$

47. $4 - 7 = 4 + (-7) = -3$

49. $-12 - 12 = -12 + (-12) = -24$

51. $-18 - (-15) = -18 + 15 = -3$

Problem Set 11.2

1. $\frac{x}{6} = 4$

$6 \bullet \frac{x}{6} = 4 \bullet 6$

$x = 24$

3. $\frac{x}{8} = -9$

$8 \bullet \frac{x}{8} = -9 \bullet 8$

$x = -72$

5. $5x = 40$

$\frac{5x}{5} = \frac{40}{5}$

$x = 8$

7. $6x = -42$

$\frac{6x}{6} = \frac{-42}{6}$

$x = -7$

9. $-11x = -77$

$\frac{-11x}{-11} = \frac{-77}{-11}$

$x = 7$

11. $-x = 78$

$\frac{-1x}{-1} = \frac{78}{-1}$

$x = -78$

13. $-3.6t = 72$

$\frac{-3.6t}{-3.6} = \frac{72}{-3.6}$

$t = -20$

15. $\frac{4}{3}x = 16$

$\frac{3}{4} \bullet \frac{4}{3}x = \frac{16}{1} \bullet \frac{3}{4}$

$x = 12$

Problem Set 11.2

17.
$$-\frac{1}{5}x = 12$$
$$-\frac{5}{1} \bullet -\frac{1}{5}x = 12 \bullet -\frac{5}{1}$$
$$x = -60$$

19.
$$\frac{1}{7} = \frac{1}{9}z$$
$$\frac{9}{1} \bullet \frac{1}{7} = \frac{1}{9}z \bullet \frac{9}{1}$$
$$\frac{9}{7} = z$$

21.
$$-\frac{4}{5}x = -\frac{7}{10}$$
$$-\frac{5}{4} \bullet -\frac{4}{5}x = -\frac{7}{10} \bullet -\frac{5}{4}$$
$$x = \frac{7}{8}$$

23.
$$-4.8z = 9.6$$
$$\frac{-4.8z}{-4.8} = \frac{9.6}{-4.8}$$
$$z = -2$$

25.
$$14 = \frac{7}{3}x$$
$$\frac{3}{7} \bullet 14 = \frac{7}{3}x \bullet \frac{3}{7}$$
$$6 = x$$

27.
$$38.3x = -306.4$$
$$\frac{38.3x}{38.3} = \frac{-306.4}{38.3}$$
$$x = -8$$

29.
$$\frac{4}{3}x = 20.08$$
$$\frac{3}{4} \bullet \frac{4}{3}x = 20.08 \bullet \frac{3}{4}$$
$$x = 15.06$$

31.
$$r = \frac{441}{7}$$
$$r = 63$$
Hoang's rate was 63 miles per hour.

33.
$$360 = 45 \bullet r$$
$$\frac{360}{45} = \frac{45r}{45}$$
$$8 = r$$
Jennifer is paid $8 an hour.

35.
$$I = \frac{110}{20} \qquad I = 5\frac{1}{2}$$
The refrigerator needs a current of $5\frac{1}{2}$ amperes.

37.
$$-7.845y = -48.639$$
$$\frac{-7.845y}{-7.845} = \frac{-48.639}{-7.845}$$
$$y = 6.2$$

39.
$$\frac{p}{-4.92} = -9.8$$
$$-4.92 \times \frac{p}{-4.92} = -9.8 \times -4.92$$
$$p = 48.216$$

41.
$$\frac{|x|}{4} = 12$$
$$4 \bullet \frac{|x|}{4} = 12.4$$
$$|x| = 48$$
$$x = 48 \text{ or } -48$$

43. $9 \bullet (-8) = -72$

45. $-64 \div (-16) = 4$

47. $-\frac{8}{5} \bullet \left(-\frac{15}{32}\right) = \frac{3}{4}$

49.
$$-4 \bullet (-9) \bullet (-8) = 36 \bullet (-8)$$
$$= -288$$

51.
$$-5 \bullet (-2) \bullet (-4) \bullet (-1)$$
$$= 10 \bullet (-4) \bullet (-1)$$
$$= -40 \bullet (-1) = 40$$

Problem Set 11.3

1.
$$7x-2 = 12$$
$$7x-2+2 = 12+2$$
$$7x = 14$$
$$\frac{7x}{7} = \frac{14}{7}; \; x = 2$$

3.
$$3x+8 = 19$$
$$3x+8+(-8) = 19+(-8)$$
$$3x = 11$$
$$\frac{3x}{3} = \frac{11}{3}; \; x = \frac{11}{3}$$

5.
$$3x-7 = -43$$
$$3x-7+7 = -43+7$$
$$3x = -36$$
$$\frac{3x}{3} = \frac{-36}{3}$$
$$x = -12$$

7.
$$-34-x = 57$$
$$34-34-x = 57+34$$
$$-x = 91$$
$$\frac{-1x}{-1} = \frac{91}{-1}$$
$$x = -91$$

Problem Set 11.3

9.
$$\frac{x}{3} - 5 = 8$$
$$\frac{x}{3} - 5 + 5 = 8 + 5$$
$$\frac{x}{3} = 13$$
$$3 \bullet \frac{x}{3} = 13 \bullet 3$$
$$x = 39$$

11.
$$-\frac{y}{2} + 8 = -3$$
$$-\frac{y}{2} + 8 + (-8) = -3 + (-8)$$
$$-\frac{y}{2} = -11$$
$$-2 \bullet -\frac{y}{2} = -11 \bullet -2$$
$$y = 22$$

13.
$$8y = 3y + 40$$
$$8y + (-3y) = 3y + 40 + (-3y)$$
$$5y = 40$$
$$\frac{5y}{5} = \frac{40}{5}$$
$$y = 8$$

15.
$$x - \frac{1}{2} = \frac{5}{2}$$
$$x - \frac{1}{2} + \frac{1}{2} = \frac{5}{2} + \frac{1}{2}$$
$$x = \frac{6}{2}$$
$$x = 3$$

17.
$$\frac{3}{5}x + \frac{1}{4} = \frac{2}{5}$$
$$20\left(\frac{3}{5}x + \frac{1}{4}\right) = \left(\frac{2}{5}\right) \bullet 20$$
$$20\left(\frac{3}{5}x\right) + 20\left(\frac{1}{4}\right) = \left(\frac{2}{5}\right) \bullet 20$$
$$12x + 5 = 8$$
$$12x + 5 + (-5) = 8 + (-5)$$
$$12x = 3$$
$$\frac{12x}{12} = \frac{3}{12}; \quad x = \frac{1}{4}$$

19.
$$5x = 3x + 8$$
$$5x + (-3x) = 3x + 8 + (-3x)$$
$$2x = 8$$
$$\frac{2x}{2} = \frac{8}{2}$$
$$x = 4$$

21.
$$8.5x - 14.2 = 16.4$$
$$8.5x - 14.2 + 14.2 = 16.4 + 14.2$$
$$8.5x = 30.6$$
$$\frac{8.5x}{8.5} = \frac{30.6}{8.5}$$
$$x = 3.6$$

23.
$$5x - 3 = 2x - 6$$
$$5x - 3 + (-2x) = 2x - 6 + (-2x)$$
$$3x - 3 = -6$$
$$3x - 3 + 3 = -6 + 3$$
$$3x = -3$$
$$\frac{3x}{3} = \frac{-3}{3}$$
$$x = -1$$

25.
$$7x + 2 = 3x - 6$$
$$7x + 2 + (-3x) = 3x - 6 + (-3x)$$
$$4x + 2 = -6$$
$$4x + 2 + (-2) = -6 + (-2)$$
$$4x = -8$$
$$\frac{4x}{4} = \frac{-8}{4}$$
$$x = -2$$

27.
$$8x + 2x = 50$$
$$10x = 50$$
$$\frac{10x}{10} = \frac{50}{10}$$
$$x = 5$$

29.
$$5y - 3y = 12$$
$$2y = 12$$
$$\frac{2y}{2} = \frac{12}{2}$$
$$y = 6$$

31.
$$8.2x + 2.4x = -106$$
$$10.6x = -106$$
$$\frac{10.6x}{10.6} = \frac{-10.6}{10.6}$$
$$x = -10$$

33.
$$6 - 5x = 4x - 7x + 18$$
$$6 - 5x = -3x + 18$$
$$6 - 5x + 5x = -3x + 18 + 5x$$
$$6 = 2x + 18$$
$$6 + (-18) = 2x + 18 + (-18)$$
$$-12 = 2x$$
$$\frac{-12}{2} = \frac{2x}{2}$$
$$-6 = x$$

35.
$$5 + 2x - 7 = 4x + 3 - x$$
$$-2 + 2x = 3x + 3$$
$$-2 + 2x + (-2x) = 3x + 3 + (-2x)$$
$$-2 = x + 3$$
$$-2 + (-3) = x + 3 + (-3)$$
$$-5 = x$$

37.
$$5y - 3 + y = 6y + 15 - 2y$$
$$6y - 3 = 4y + 15$$
$$6y - 3 + (-4y) = 4y + 15 + (-4y)$$
$$2y - 3 = 15$$
$$2y - 3 + 3 = 15 + 3$$
$$2y = 18$$
$$\frac{2y}{2} = \frac{18}{2}$$
$$y = 9$$

Problem Set 11.3

39. $\frac{3}{2}x+\frac{1}{2}x = 4x+\frac{5}{2}+\frac{3}{2}x$

$2\left(\frac{3}{2}x+\frac{1}{2}x\right) = 2\left(4x+\frac{5}{2}+\frac{3}{2}x\right)$

$2\left(\frac{3}{2}x\right)+2\left(\frac{1}{2}x\right)$

$= 2(4x)+2\left(\frac{5}{2}\right)+2\left(\frac{3}{2}x\right)$

$3x+x = 8x+5+3x$

$4x = 11x+5$

$4x+(-11x) = 11x+5+(-11x)$

$-7x = 5$

$\frac{-7x}{-7} = \frac{5}{-7}$

$x = -\frac{5}{7}$

41. $14,000 = V-3250$

$14,000+3250 = V-3250+3250$

$17,250 = V$

The original value of the truck is $17,250.

43. $375 = V_0+32 \cdot 4$

$375 = V_0+128$

$375+(-128) = V_0+128+(-128)$

$247 = V_0$

The initial velocity is 247 feet per second.

45. $2200 = 50 \cdot 19+B$

$2200 = 950+B$

$2200+(-950) = 950+B+(-950)$

$1250 = B$

Jean's base salary was $1250.

47. $A = \frac{4 \text{ in.} \cdot 3 \text{ in.}+4 \text{ in.} \cdot 5 \text{ in.}}{2}$

$A = \frac{12 \text{ in.}^2+20 \text{ in.}^2}{2}$

$A = \frac{32 \text{ in.}^2}{2}$

$A = 16 \text{ in.}^2$

The area of the trapezoid is 16 square inches.

49. addition

51. $-3.256x+42.38 = -16.8792$

$-3.256x+42.38-42.38$

$= -16.8792-42.38$

$-3.256x = -59.2592$

$\frac{-3.256x}{-3.256} = \frac{-59.2592}{-3.256}$

$x = 18.2$

53. $8\frac{1}{4}x-5x = 5\frac{1}{8}-\frac{1}{2}x$

$\frac{33}{4}x-5x = \frac{41}{8}-\frac{1}{2}x$

$8\left(\frac{33}{4}x-5x\right) = 8\left(\frac{41}{8}-\frac{1}{2}x\right)$

$8\left(\frac{33}{3}x\right)-8(5x) = 8\left(\frac{41}{8}\right)-8\left(\frac{1}{2}x\right)$

$66x-40x = 41-4x$

$26x = 41-4x$

$26x+4x = 41-4x+4x$

$30x = 41$

$\frac{30x}{30} = \frac{41}{30}$

$x = \frac{41}{30}$

55. $-(-8) = 8$

57. $-(7.4) = -7.4$

59. $0 > -8$

61. $6 > -7$

63. $|-8| = 8$

65. $|17.5| = 17.5$

Problem Set 11.4

1. $x + 6$

3. $x - 10$

5. $3x$

7. $x + 3$

9. $x - 8$

11. $3x+5 = 23$

$3x+5-5 = 23-5$

$3x = 18$

$\frac{3x}{3} = \frac{18}{3}$; $x = 6$

13. $4x-14 = 94$

$4x-14+14 = 94+14$

$4x = 108$

$\frac{4x}{4} = \frac{108}{4}$; $x = 27$

Problem Set 11.4

15. $\frac{x}{2}+\frac{5}{8} = 2x$
$8\left(\frac{x}{2}+\frac{5}{8}\right) = 8(2x)$
$4x+5 = 16x$
$4x-4x+5 = 16x-4x$
$5 = 12x$
$\frac{5}{12} = \frac{12x}{12};\ \frac{5}{12} = x$

17. electricity $= 2x = 2 \bullet 20 = 40$
dry cleaning $= x = 20$
food $= 3 \bullet 2x = 6x = 6 \bullet 20$
$= 120$
$180 = 2x+x+6x$
$180 = 9x$
$\frac{180}{9} = \frac{9x}{9};\ 20 = x$
Mrs. Gonzales spends \$40 on electricity, \$20 on dry cleaning, and \$120 on food.

19. width $= w = 16$ m
length $= 4w = 4 \bullet 16$ m $= 64$ m
160 m $= 2 \bullet w+2 \bullet 4w$
160 m $= 2w+2 \bullet 4w$
160 m $= 10w$
$\frac{160 \text{ m}}{10} = \frac{10w}{10};\ 16 \text{ m} = w$
The width is 16 m and the length is 64 m.

21. width $= w = 22$
length $= 18+w = 18+22 = 40$
$124 = 2 \bullet w+2 \bullet (18+w)$
$124 = 2w+36+2w$
$124-36 = 4w$
$88 = 4w$
$\frac{88}{4} = \frac{4w}{4};\ 22 \text{ m} = w$
Kevin should make his garden 22 feet by 40 feet.

23. Price of dress before reduction $= p$
$35.16 = p-0.40 \times p$
$35.16 = (1-0.40) \times p$
$35.16 = 0.60 \times p$
$\frac{35.16}{0.60} = \frac{0.60 \times p}{0.60}$
$58.6 = p$
The price of the dress before the reduction was \$58.60.

25. Salary before increase $= s$
$17{,}280 = s+0.08 \times s$
$17{,}280 = (1+0.08) \times s$
$17{,}280 = 1.08 \times s$
$\frac{17{,}280}{1.08} = \frac{1.08 \times s}{1.08}$
$16{,}000 = s$
John's salary before the increase was \$16,000.

27. Bushels of wheat harvested last year $= b$
$34{,}200 = 1.20 \times b$
$\frac{34{,}200}{1.20} = \frac{1.20 \times b}{1.20}$
$28{,}500 = b$
The farmer harvested 28,500 bushels of wheat last year.

29. sales $= s$
$5100 = 700+0.088 \times s$
$5100-700 = 700+0.088 \times s-700$
$4400 = 0.088 \times s$
$\frac{4400}{0.088} = \frac{0.088 \times s}{0.088}$
$50{,}000 = s$
Neil's sales were \$50,000 for April.

31. $-4(3x-2) = -4 \bullet 3x-(-4) \bullet 2$
$= -12x+8$

33. 9a, -6b

35. $8x+2x = (8+2)x$ or $(4+1)2x$

37. $6x+x = 7x$

39. $5y-4.2y = 0.8y$

41. $9x+2y-4x-6y$
$= 9x-4x+2y-6y$
$= 5x-4y$

Chapter 11 Additional Exercises

1. $x-9 = 17$
$x-9+9 = 17+9$
$x = 26$

3. $x-\frac{3}{5} = 2$
$x-\frac{3}{5}+\frac{3}{5} = 2+\frac{3}{5}$
$x = 2\frac{3}{5}$

Chapter 11 Additional Exercises

5.
$$-12 = y-8$$
$$-12+8 = y-8+8$$
$$-4 = y$$

7.
$$\frac{x}{9} = 8$$
$$9 \bullet \frac{x}{9} = 8 \bullet 9$$
$$x = 72$$

9.
$$-5x = 32$$
$$\frac{-5x}{-5} = \frac{32}{-5}$$
$$x = -\frac{32}{5}$$

11.
$$\frac{3}{4}x = -\frac{8}{7}$$
$$\frac{4}{3} \bullet \frac{3}{4}x = -\frac{8}{7} \bullet \frac{4}{3}$$
$$x = -\frac{32}{21}$$

13.
$$-31.4x = 188.4$$
$$\frac{-31.4x}{-31.4} = \frac{188.4}{-31.4}$$
$$x = -6$$

15.
$$\frac{|x|}{16} = 4$$
$$16 \bullet \frac{|x|}{16} = 4 \bullet 16$$
$$|x| = 64$$
$$x = \pm 64$$

17.
$$8x-6 = 18$$
$$8x-6+6 = 18+6$$
$$8x = 24$$
$$\frac{8x}{8} = \frac{24}{8}$$
$$x = 3$$

19.
$$8x-3 = 6x-5$$
$$8x-6x-3 = 6x-5-6x$$
$$2x-3 = -5$$
$$2x-3+3 = -5+3$$
$$2x = -2$$
$$\frac{2x}{2} = \frac{-2}{2}$$
$$x = -1$$

21.
$$3+6x-4 = 9x-12+8x$$
$$6x-1 = 17x-12$$
$$6x-1-17x = 17x-12-17x$$
$$-11x-1 = -12$$
$$-11x-1+1 = -12+1$$
$$-11x = -11$$
$$\frac{-11x}{-11} = \frac{-11}{-11}$$
$$x = 1$$

23.
$$7\frac{1}{3}x-4 = 6\frac{1}{8}+3\frac{1}{4}x$$
$$\frac{22}{3}x-4 = \frac{49}{8}+\frac{13}{4}x$$
$$24\left(\frac{22}{3}x-4\right) = 24\left(\frac{49}{8}+\frac{13}{4}x\right)$$
$$176x-96 = 147+78x$$
$$176x-96-78x = 147+78x-78x$$
$$98x-96 = 147$$
$$98x-96+96 = 147+96$$
$$98x = 243$$
$$\frac{98x}{98} = \frac{243}{98}$$
$$x = 2\frac{47}{98}$$

25.
$$3x-12 = 66$$
$$3x-12+12 = 66+12$$
$$3x = 78$$
$$\frac{3x}{3} = \frac{78}{3}$$
$$x = 26$$

27.
length = l
width = $l - 28$
$200 = 2 \bullet l + 2 \bullet (l - 28)$
$200 = 2l + 2l - 56$
$200 = 4l - 56$
$200 + 56 = 4l - 56 + 56$
$256 = 4l$
$\frac{256}{4} = l;\ 64 = l$
The length of the rectangle is 64 m and the width of the rectangle is 36 m.

29.
Monthly payments = p
$\frac{16{,}000-2500}{45} = p;\ 300 = p$
Aaron's monthly payments are \$300.

Chapter 11 Practice Test

1.
$$x-9 = 21$$
$$x-9+9 = 21+9$$
$$x = 30$$

3.
$$5-6y = -9-5y$$
$$5-6y+5y = -9+5y+5y$$
$$5-y = -9$$
$$5-y = -9-5$$
$$-y = -14$$
$$\frac{-y}{-1} = \frac{-14}{-1}$$
$$y = 14$$

5.
$$x+8 = 27$$
$$x+8-8 = 27-8$$
$$x = 19$$

Chapter 11 Practice Test

7. $6x-4 = 7x+8$
$6x-4-6x = 7x+8-6x$
$-4 = x+8$
$-4-8 = x+8-8$
$-12 = x$

9. $423.8y-49.6 = 162.3$
$423.8y-49.6+49.6 = 162.3+49.6$
$423.8y = 211.9$
$\frac{423.8y}{423.8} = \frac{211.9}{423.8}$
$y = 0.5$

11. $\frac{x}{3}+\frac{3}{4} = 2x$
$12\left(\frac{x}{3}+\frac{3}{4}\right) = 2x \bullet 12$
$4x+9 = 24x$
$4x+9-4x = 24x-4x$
$9 = 20x$
$\frac{9}{20} = \frac{20x}{20}$
$\frac{9}{20} = x$

13. Original price = p
$7840 = p-0.20p$
$7840 = (1-0.20)p$
$7840 = 0.80p$
$\frac{7840}{0.80} = \frac{0.80p}{0.80}$
$9800 = p$
The original price was $9800.

Problem Set 12.1

1. Rent is the greatest expense in the budget.

3. $\frac{\$240}{\$2400} = \frac{1}{10}$

5. $\frac{\$300}{\$600} = \frac{1}{2}$

7. $30,000 + 1000 + 19,000 + 20,000 + 80,000 + 50,000 = 200,000$
The total number of viewers is 200,000.

9. $\frac{80,000 \text{ viewers}}{200,000 \text{ viewers}} = \frac{2}{5}$

11. $\frac{19,000 \text{ viewers}}{30,000 \text{ viewers}} = \frac{19}{30}$

13. $a = 3\% \bullet \$50,000,000$
$a = 0.03 \bullet \$50,000,000$
$a = \$1,500,000$

15. $a = 12\% \bullet \$50,000,000$
$a = 0.12 \bullet \$50,000,000$
$a = \$6,000,000$

17. $a = 6\% \bullet \$50,000,000$
$a = 0.06 \bullet \$50,000,000$
$a = \$3,000,000$

19. $a = 15\% \bullet \$180,000$
$a = 0.15 \bullet \$180,000$
$a = \$27,000$

21. $a = 35\% \bullet \$180,000$
$a = 0.35 \bullet \$180,000$
$a = \$63,000$

23. $a = 12\% \bullet \$180,000$
$a = 0.12 \bullet \$180,000$
$a = \$21,600$

25. Numerical information or data.

27. 100

29. $a = 32.4\% \bullet \$48,000,000$
$a = 0.324 \bullet \$48,000,000$
$a = \$15,552,000$

31. $7+y = 16$
$7+y-7 = 16-7$
$y = 9$

33. $x-7 = -15$
$x-7+7 = -15+7$
$x = -8$

35. $6.8+y = -3.2$
$6.8+y-6.8 = -3.2-6.8$
$y = -10$

37. $p+\frac{3}{8} = \frac{4}{7}$
$p+\frac{3}{8}-\frac{3}{8} = \frac{4}{7}-\frac{3}{8}$
$p = \frac{32}{56}-\frac{21}{56}$
$p = \frac{11}{56}$

Problem Set 12.1

39.
$$-\frac{3}{5} = p - \frac{1}{15}$$
$$-\frac{3}{5} + \frac{1}{15} = p - \frac{1}{15} + \frac{1}{15}$$
$$-\frac{9}{15} + \frac{1}{15} = p$$
$$-\frac{8}{15} = p$$

43.
$$\frac{y}{7} = -8$$
$$7 \bullet \frac{y}{7} = -8 \bullet 7$$
$$y = -56$$

45.
$$-x = -34$$
$$\frac{-x}{-1} = \frac{-34}{-1}$$
$$x = 34$$

47.
$$-4.3y = -35.26$$
$$\frac{-4.3y}{-4.3} = \frac{-35.26}{-4.3}$$
$$y = 8.2$$

49.
$$-8x = -34.4$$
$$\frac{-8x}{-8} = \frac{-34.4}{-8}$$
$$x = 4.3$$

51.
$$\frac{3}{5}x = \frac{9}{10}$$
$$\frac{5}{3} \bullet \frac{3}{5}x = \frac{9}{10} \bullet \frac{5}{3}$$
$$x = \frac{3}{2}$$

Problem Set 12.2

1. 3500 students
3. 4500 students
5. year 2000
7. 20,000 people
9. Friday, 1993; 35,000 people
11. 35,000 - 30,000 = 5000 people
13. 16,000 cars
15. 23,000 - 18,000 = 5000 cars
17. March
19. $100,000
21. 1988; $70,000
23. $120,000 = $70,000 = $50,000
25. 65° F
27. 55° F
29. 80° F - 50° F = 30° F
31. 2000 hamburgers
33. June, 1993; 6000 hamburgers
35. 5000 - 2500 = 2500 hamburgers
37. double
39. comparison
41. 4000 + 2000 + 2000 + 4000 + 5000 + 6000 = 23,000

Problem Set 12.3

1. 12 years
3. 22 to 24 inches
5. 7 + 12 + 11 + 8 = 38 years
7. 10 customers
9. $0 to $15
11. 10 + 15 + 12 + 6 + 4 = 47 customers
13. 14 students
15. 20 to 25 years of age
17. 10 + 14 + 12 + 6 = 42 students
19. 10 + 14 + 12 + 6 + 4 + 5 + 2 + 1 + 2 + 1 = 57
21. 18 employees
23. 14 + 12 + 6 = 32 employees
25. 8 + 18 = 26 employees

Problem Set 12.3

27. $\frac{8 \text{ employees}}{74 \text{ employees}} = \frac{4}{37}$

29.
$9x+4 = 85$
$9x+4-4 = 85-4$
$9x = 81$
$\frac{9x}{9} = \frac{81}{9}$
$x = 9$

31.
$6.3x+7.2 = -40.05$
$6.3x+7.2-7.2 = -40.05-7.2$
$6.3x = -47.25$
$\frac{6.3x}{6.3} = \frac{-47.25}{6.3}$
$x = -7.5$

33.
$6x-7 = 3x-43$
$6x-3x-7 = 3x-43-3x$
$3x-7 = -43$
$3x-7+7 = -43+7$
$3x=-36$
$\frac{3x}{3} = \frac{-36}{3}$
$x = -12$

35.
$3-2x = 5x-8x+15$
$3-2x = -3x+15$
$3-2x+3x = -3x+15+3x$
$3+x = 15$
$3+x-3 = 15-3$
$x = 12$

37.
$8y-2+y = 5y+14-3y$
$9y-2 = 2y+14$
$9y-2y-2 = 2y+14-2y$
$7y-2 = 14$
$7y-2+2 = 14+2$
$7y = 16$
$y = \frac{16}{7}$

Problem Set 12.4

1. $\frac{76+83+97+42+65}{5} = 72.6$

3. $\frac{7800+6450+5870+7200}{4} = 6830$

5. $\frac{3.7+5.6+8.9+7.4}{6} + \frac{6.2+4.1}{6} = 5.98\overline{3}$

7. 28, 34, 36, 44, 45
The median is 36.

9. 17, 18, 24, 28, 30, 32
$\frac{24+28}{2} = 26$
The median is 26.

11. 104, 108, 132, 160, 210, 314
$\frac{132+160}{2} = 146$
The median is 146.

13. 15.2, 18.3, 19.6, 25.8, 36.7
The median is 19.6.

15. 2.4, 3.2, 4.8, 6.3, 8.7, 9.1
$\frac{4.8+6.3}{2} = 5.55$
The median is 5.55.

17. The mode is 3.

19. There is no mode.

21. The modes are 72 and 85.

23. The mode is 14.

25. $\frac{78+84+86+93+95}{5} = 87.2$

27. $\frac{\$1.99+\$1.50}{4} + \frac{\$2.05+\$2.25}{4} \approx \$1.95$

29. $\frac{3.6 \text{ in.}+4.2 \text{ in.}}{4} + \frac{3.8 \text{ in.}+4.2 \text{ in.}}{4} = 3.95 \text{ in.}$

31. \$80,000; \$95,000; \$115,000; \$130,000; \$150,000
The median is \$115,000.

33. 12, 17, 24, 32
$\frac{17+24}{2} = 20.5$
The median is 20.5.

35. $5\frac{1}{8}, 6\frac{1}{2}, 7\frac{3}{4}, 8\frac{1}{4}$

$$\frac{6\frac{1}{2}+7\frac{3}{4}}{2}=7\frac{1}{8}$$

The median is $7\frac{1}{8}$ hours.

37. The mode is 28.

39. sum

41. $\frac{15.3+18.2+17.1}{9}+\frac{16.4+20.8+14.7}{9}+\frac{17.5+16.8+20.4}{9}$

≈ 17.5

The mean is about 17.5 inches.

43. The mode is 96.3.

45. $2x$

47. $x - 20$

49. Laura's earnings = x
Jean's earnings = $3x$
$x+3x = 544$
$4x = 544$
$\frac{4x}{4} = \frac{544}{4}$
$x = 136$
$3x = 408$
Jean earns $408 per week.

51. Original price = p
$77.40 = p-0.40p$
$77.40 = (1-0.40)p$
$77.40 = 0.60p$
$\frac{77.40}{0.60} = \frac{0.60p}{0.60}$
$129 = p$
The original price was $129.

53. length = l
width = $l - 16$
$120 = 2l+2(l-16)$
$120 = 2l+2l-32$
$120 = 4l-32$
$120+32 = 4l-32+32$
$152 = 4l$
$\frac{152}{4} = \frac{4l}{4}$
$38 = l$
The fence should be 38 feet long and 22 feet wide.

Chapter 12 Additional Exercises

1. $1250 + $200 + $1000 + $50 + $500 = $3000

3. $\frac{\$50}{\$3000} = \frac{1}{60}$

5. $\frac{\$200}{\$1250} = \frac{4}{25}$

7. $a = 6\% \bullet 24$
$a = 0.06 \bullet 24$
$a = 1.44$

9. $a = 15\% \bullet 24$
$a = 0.15 \bullet 24$
$a = 3.6$

11. $a = 16\% \bullet 24$
$a = 0.16 \bullet 24$
$a = 3.8$

13. 8,000

15. 8,000 - 6,000 = 2,000

17. August

19. 5,000 + 6,000 + 4,000 + 6,500 + 8,000 + 9,000 = 38,500

21. $90,000

23. May, 1993; $80,000

25. February;
$110,000 - $90,000 = $20,000

27. 4 years

29. 4 + 2 = 6 years

31. 7 + 4 + 2 = 13 years

33. 500 employees

35. 60 - 65 years of age

37. 3,000 + 5,000 + 3,000 + 2,000 + 1,000 + 500 = 14,500

39. $\frac{8+12+9+14+7+22+16}{7} \approx 12.6$

Chapter 12 Additional Exercises

41. $\frac{1200+1400+1700+2000}{4} = 1575$

43. 7, 8, 9, 12, 14, 16, 22
The median is 12.

45. 1200, 14,00, 1700, 2000
$\frac{1400+1700}{2} = 1550$
The median is 1550.

47. The mode is 6.

49. There is no mode.

51. $\frac{\$13{,}999+\$12{,}850}{4} + \frac{\$14{,}500+\$11{,}300}{4}$
$= \$13{,}162.25$

53. 215.2, 216.5, 229.1, 244.2, 252, 266.3
$\frac{229.1+244.2}{2} = 236.65$
The median is 236.65.

55. The mode is \$0.99.

Chapter 12 Practice Test

1. $a = 3\% \bullet \$3500$
$a = 0.03 \bullet \$3500$
$a = \$105$

3. $a = 17\% \bullet \$3500$
$a = 0.17 \bullet \$3500$
$a = \$595$

5. 1990

7. 50,000

9. 10,000 + 30,000 + 45,000
= 85,000

11. 940, 1100, 1450, 1680
$\frac{1100+1450}{2} = 1275$

13. $\frac{\$15.98+\$21.05}{4} + \frac{\$36.17+18.32}{4} = \$22.88$

Final Examination

1. Two hundred ten thousand, twenty-four

3.
$$\begin{array}{r} \;12\,9 \\ 7\;\cancel{2}\;1\cancel{0}15 \\ \cancel{8}\;\cancel{3}\;\cancel{0}\;\cancel{5} \\ -\underline{7\;6\;4\;9} \\ 6\;5\;6 \end{array}$$

5.
$$\begin{array}{r} 10 \\ 92\overline{)967} \\ \underline{92} \\ 47 \end{array}$$
The answer is 10 R 47.

7. Does $9 \bullet 16 = 4 \bullet 37$?
Since $144 \neq 148$,
$\frac{9}{4} \neq \frac{37}{16}$

9.
$$\begin{aligned} \frac{15}{7} + 45 &= \frac{15}{7} \div \frac{45}{1} \\ &= \frac{15}{7} \bullet \frac{1}{45} \\ &= \frac{1\cancel{5}}{7} \bullet \frac{1}{3 \bullet 1\cancel{5}} \\ &= \frac{1}{21} \end{aligned}$$

11.
$$\begin{aligned} \frac{9}{7} + \frac{5}{3} &= \frac{9}{7} \bullet \frac{3}{3} + \frac{5}{3} \bullet \frac{7}{7} \\ &= \frac{27}{21} + \frac{35}{21} \\ &= \frac{62}{21} \end{aligned}$$

13.
$$\begin{aligned} \frac{9}{5} - \frac{3}{10} &= \frac{9}{5} \bullet \frac{2}{2} - \frac{3}{10} \\ &= \frac{18}{10} - \frac{3}{10} \\ &= \frac{15}{10} \\ &= \frac{3}{2} \end{aligned}$$

15.
$$\begin{aligned} 5\frac{5}{8} \bullet 3\frac{1}{9} &= \frac{45}{8} \bullet \frac{28}{9} \\ &= \frac{5 \bullet \cancel{9}}{2 \bullet \cancel{4}} = \frac{\cancel{4} \bullet 7}{\cancel{9}} \\ &= \frac{35}{2};\ = 17\frac{1}{2} \end{aligned}$$

17. $\frac{17}{22} < \frac{7}{8}$, since $\frac{68}{88} < \frac{77}{88}$

Final Examination

19.
$$\begin{array}{r} 307 \\ 6.48 \\ +\underline{\;\;0.395} \\ 313.875 \end{array}$$

21.
$$\begin{array}{r} 9.84 \\ \times\underline{8.70} \\ 6\;8880 \\ \underline{78\;72\;\;\;\;} \\ 85.6080 \end{array}$$
The cost is about $85.61.

23. $\dfrac{6.7+24.8+41+36.3}{4} = 27.2$

25. $\dfrac{720 \text{ feet}}{80 \text{ seconds}} = 9\dfrac{\text{feet}}{\text{second}}$

27. Does $27 \bullet 105 = 63 \bullet 45$?
Since $2835 = 2835$, then the proportion is true.

29. $7.5\% = 0.075$

31.
$51 = p\% \times 170$
$51 = p \times 0.01 \times 170$
$51 = p \times 1.7$
$\dfrac{51}{1.7} = \dfrac{p \times 1.7}{1.7}$
$30 = p$

33. Amount of increase
$= \$32{,}000 - \$30{,}000$
$= \$2000$
Percent increase
$= \dfrac{2000 \bullet 100}{30{,}000} \approx 6.7$

35. $5^3 = 5 \times 5 \times 5 = 125$

37. $\sqrt{169} = 13$

39.
$$\begin{array}{rl} \overset{74}{\not{7}\not{5}} \text{ minutes} & \overset{90}{\not{3}\not{0}} \text{ seconds} \\ -\underline{29 \text{ minutes}} & \underline{54 \text{ seconds}} \\ 45 \text{ minutes} & 36 \text{ seconds} \end{array}$$

41. $21.5 \text{ km} = 21{,}500 \text{ m}$

43. $F = \dfrac{9}{5} \bullet 45 + 32 = 113$

45. $A = 17 \text{ ft} \bullet 25 \text{ ft} = 425 \text{ ft}^2$

47. $C \approx 3.14 \times 18 \text{ in} \approx 56.52 \text{ in}$

49. $V = 4 \text{ yd} \times 7 \text{ yd} \times 8.9 \text{ yd}$
$= 249.2 \text{ yd}^3$

51. $16-(-23) = 16+23 = 39$

53.
$4x-6y-9x-5y+4$
$4x-9x-6y-5y+4$
$-5x-11y+4$

55.
$14-3y = -22+6y-8y$
$14-3y = -22-2y$
$14-3y+22 = -22-2y+22$
$36-3y = -2y$
$36-3y+3y = -2y+3y$
$36 = y$

57. width $= w$
length $= 3w$
$144 = 2w+2 \bullet 3w$
$144 = 2w+6w$
$144 = 8w$
$\dfrac{144}{8} = \dfrac{8w}{8}$
$18 = w$
The width is 18 m and the length is 54 m.

59. $7000 - $2000 = $5000